Introduction

When Britain's Avro Lancaster undertook its first operational sorties in March 1942 it quickly established itself as the supreme RAF heavy bomber of its day and at the end of the Second World War, in Europe at least, it could still lay claim to that title despite the abilities of the radial-engined Handley Page Halifax. In fact it was only with the later arrival of the somewhat more advanced Avro Lincoln (initially developed as the Lancaster B.IV and B.V) and the introduction to Europe of the Boeing B-29 Superfortress which served to tip the Lancaster from its perch – and that didn't happen immediately. In fact it wasn't until 1950 that RAF Bomber Command was finally able to dispose of its last operational Lancaster bombers thereby dispelling the vague notion (which still lingers in some quarters) that the Lancaster must have been declared obsolete the day after the war in Europe finished and immediately scrapped. In truth of course the Lancaster carried on to find useful employment with the RAF until October 1956, while abroad the Lancaster was destined to serve on, in diminishing numbers until 1965 – a full twenty years after the Second World War had finished!

The object of this book is to illustrate as far as possible, within the constraints of a single volume, using archive images, specially commissioned colour illustrations and photographs of completed models, the many 'varieties' of Lancaster that once existed in the post-war era. It is hoped that this work will open a door for those modellers who wish to complete their kit in a different 'hue' to the standard, yet very familiar wartime RAF 'Lanc'. Although this book does not attempt to repeat the developmental, technical and wartime history of this aircraft – so often achieved by others in the past – it does include an explanatory text, supported by extended photo-captions, summarising the Lancaster's service with the Royal Air Force, Royal Canadian Air Force and *Aéronavale* with brief references to its service with Argentina, Sweden and Egypt. It is hoped that the non-modelling but nostalgic fraternity might therefore find this book to be of interest too.

Acknowledgements

The authors would like to express their gratitude to Mike Smith, Rosalyn Blackmore and all the staff at Newark Air Museum who yet again opened their archive and museum facilities to us with unfailing kindness. Grateful thanks are also extended to Roger Lindsay, Tony O'Toole, Steve Hague, Brandon White and Doug Derry for their invaluable contributions. Thanks are also due to the members of the IPMS (UK) Bomber Command Special Interest Group and to Mike Verier for the loan of their models with additional thanks being extended to Bob Holler who photographed them. Last, and by no means least, our special thanks too to Mark Gauntlet for his superb artwork.

Martin Derry and Neil Robinson

NOTES

1. During 1947 the Roman numerals that had previously been used to represent RAF Mark numbers were replaced by Arabic numerals, thus the Lancaster B.VII, for instance, became the B.7. Irrespective of other considerations Arabic-style numerals are employed when referring to Marks immediately following VE-Day in May 1945; they are used for simplicity's sake and to avoid unnecessary duplication given that the Lancaster served through the transitional period.

2. B for Bomber: During 1942 the prefix 'B' began to appear with reference to RAF bombers whereby Lancaster Mks.I and III became the Lancaster B.I and B.III. For the sake of continuity (excluding early to mid-wartime bombers) the designator B (Bomber) is used throughout.

RCAF Lancaster 10MP KB889 following retirement.
Author's collection

Origins

A twin-engined Avro Manchester Ia (by virtue of its two enlarged fins as opposed to the three smaller fins of the Manchester I). Its influence upon the later Lancaster is not to be denied.
Author's collection

The Lancaster was developed from its immediate predecessor the Avro Manchester, a twin-engined bomber powered by two Rolls-Royce 24-cylinder Vulture engines which, it was hoped, would deliver and maintain a little under 1,800hp each. Today it is fairly common knowledge that while the Manchester's airframe showed great promise – the engines did not, something that became evident during the prototype's (L7246) first flight on 25 July 1939 when both engines ran hot and were unable to develop their predicted power output. Initial problems experienced with the airframe were largely overcome by increasing the Manchester's wingspan to 90ft 1in and by the provision of a third vertical tail fin to cure a lack of directional stability and which externally served to distinguish a Manchester I from the later Mk.Ia – the latter having all three fins deleted and replaced by two significantly larger fins and rudders. There was no swift cure for the Vulture's problems however, and just 200 production Manchesters were completed.

Today, in retrospect, it is immediately evident that the Manchester Ia possessed the proportional look of a (two-engined) Lancaster albeit with a Fraser-Nash FN.7A dorsal turret. However, Avro's chief designer, Roy Chadwick – the man who designed the Manchester – was aware that the airframe and its capacious uninterrupted bomb bay possessed significant growth potential and had instigated work on a four-engined Manchester long *before* the Manchester I even entered squadron service. The Company was ready to progress with a prototype in late 1940 and significantly, in addition to making the necessary changes to enlarge the basic airframe, they replaced the two Vultures with four Rolls-Royce Merlin engines. Temporarily referred to as the Manchester III for security purposes the prototype Lancaster, BT308, first flew on 9 January 1941, a bare six weeks after work on it had commenced. Ultimately the name 'Lancaster' was bestowed upon the new design which also served to disassociate it from its unfortunate forebear.

Between undertaking its first mission late on the evening of 3 March 1942, when just four Lancasters laid mines in German waters, and late April 1945, Lancasters completed in excess of 152,000 operational sorties which resulted in the loss of approximately 3,460 of their number, *excluding* those lost in wartime accidents or written-off as a result of battle-damage. (A further 93 RAF Lancasters, of all Marks, were lost between 1946 and 1956 [including Canadian-built B.10, KB705 in March 1946]. The first was B.3, LM649, which damaged a wing tip in a hangar on 3 January 1946 and was struck off charge (SOC), while the last was MR.3, RE211, on 2 February 1956 which was SOC following an aborted takeoff run.)

To some the least successful of Bomber Command's wartime *four*-engined bombers was the surprisingly manoeuvrable Short Stirling which, despite shortcomings, continued to be employed as a strategic bomber until September 1944. 'OJ-R' is N6124 from No.149 Squadron which arrived with the unit in December 1941 and crashed in France on 5 May 1942.
Author's collection

A Merlin-engined Handley Page Halifax II. Although the Halifax airframe would develop into a robust bomber, its marriage to the Merlin engine was less successful than originally predicted. Later, a more developed airframe would be mated with Bristol Hercules radial engines to produce a more satisfactory aircraft and bomber. W1245 'EY-B', seen here, belonged to No.78 Squadron with which it served for about four weeks before it was shot down over Belgium on 11/12 August 1942. W1245's mid-upper turret is masked by the aircraft's wing. *Author's collection*

Despite its qualities the radial-engined Halifax had to be sacrificed by Bomber Command at war's end and all were immediately transferred to Transport Command or other units. The variant seen here is a Halifax A.IX, a supply and troop-carrying aircraft developed directly from the Halifax B.VII with which it was virtually identical except for its reduced armament.
Author's collection

Sources conflict as to precisely how many Lancasters were built, but the authors believe that the figure stands at 7,377 (including three prototypes) with production ceasing in 1946.

As the war in Europe neared its end, the RAF, in common with the rest of Britain's armed forces, was ordered to prepare for a post-war reduction in size, the country's economy being, as ever, in a parlous state with utility and thrift as the watchwords of the day. Effectively bankrupt, the British Government was ultimately forced to negotiate with the USA in December 1945 for a

As an insurance against a possible shortfall in Merlin engine production, 300 Lancasters were fitted with Bristol Hercules radial engines to become the Lancaster B.II, although in the event a serious shortage of Merlin engines never arose. B.II, DS689 'OW-S', of No.426 Squadron RCAF was shot down on 7/8 October 1943. *Author's collection*

The definitive wartime Lancaster – barring unit codes of course – seen during its delivery flight to No.38 MU at the very end of the war. B.III, RE172, remained with MUs until it was delivered to No.37 Squadron in the Middle East in April 1946. It was SOC in June 1947. *Author's collection*

loan of just over £1 billion – a colossal sum that was to be repaid in fifty annual instalments from 1950 (the final payment was made in 2006).

Following VE-Day, Bomber Command was required to operate just one type of four-engined bomber until later designs became available. The Lancaster was chosen and as early as 7 May 1945 the axe fell upon the Command's Halifax fleet when all eleven

squadrons of No.4 Group were transferred to Transport Command with any others being rapidly allocated elsewhere, stored, sold or scrapped. As an RAF bomber the Halifax was finished (the Stirling having ceased strategic bombing operations in 1944) – for the Lancaster a future with Bomber Command remained.

Of course in May 1945 the war against Japan was still a continuing fact and it was envisaged that Britain would deploy as many resources as possible to that theatre – which was to include tropicalized Lancasters (amongst other types) under the auspices of 'Tiger Force'; however, the dropping of the two atomic bombs on Japan in August 1945 brought that conflict to an unexpectedly abrupt end with 'Tiger Force' being eventually laid to rest on 31October 1945. With the demise of that Force its specially adapted Lancaster FE (Far East) variants, distinguished by their all-white upper surfaces and black undersides, became available for deployment elsewhere.

For Coastal Command, the ending of the wartime Lend-Lease Act determined that all surviving aircraft and equipment supplied under that Act must either be returned to the USA or paid for – there was no ambiguity – which meant that the Command would ultimately lose their existing long-range maritime patrol aircraft as supplied by the US Government under the terms of the Act. That in turn would leave a significant gap in Maritime Britain's defences; defences which until 2010 even politicians regarded as essential! Given its considerable endurance and load-carrying capability a suitably adapted Lancaster could usefully plug the gap until a purpose-designed aircraft was procured.

Variations on a Theme
A curiosity of the wartime Lancaster, as distinct from many other types of warplane, was that the *primary* production examples: B.I, B.III, B.VII and B.X, in terms of their major components, were all essentially the same. To be sure steady improvements were made but in this case the increase in Mark number did not indicate a commensurate leap in capability or overall performance, while those of the Spitfire, for instance, did; as the latter progressed through the Marks so did its performance – significantly. By comparison, the very last Lancasters delivered, in March 1946, were still Mk.1s albeit benefiting from most of the improvements that had been developed and introduced as the war went on.

That differences existed between Marks isn't disputed, it is merely that they were not necessarily apparent *externally*; a case in point being whether Rolls-Royce or American-supplied Packard-built Merlin engines were fitted. The shape and positioning of the dorsal turret could help identify a certain Mark assuming it had been retained, but in the post-war period they were frequently removed if only to allow aircrew to move about the aircraft's fuselage more readily;

internally at least, dorsal turrets were very cumbersome. When retained, both the B.7 and Canadian B.10 employed an American Glenn Martin turret with twin-Browning .5in machine-guns (mgs) which were sited further forward along the fuselage than other Marks and did not require a fairing. The B.1 and B.3 featured dorsal turrets with twin-Browning .303in mgs and a distinctive fairing. Both the B.1(FE) and B.7(FE) received Fraser Nash FN.82 rear turrets with twin .5in mgs. Surviving B.3s which were converted to maritime ASR.3, GR.3 or MR.3 standard gained observation panels on both sides of the fuselage forward of the tailplane. They lost their dorsal turrets but retained their four-gun tail and two-gun nose turrets which, though normally unarmed, could be rearmed as required.

Such distinctions, although useful, were therefore only valid while each Lancaster retained its dorsal turret – or indeed, any guns at all: in which case, viewed externally and ignoring additional observation panels if fitted, they could all be equally described as 'Mark 1s'. An appropriate illustration of this is provided by the B.7 of which 180 were built and delivered between April 1945 and January 1946. However, fifty others had been constructed a little earlier with completion delayed because their Martin dorsal turrets had failed to arrive. Occasionally referred to as the B.7 'Interim' each received instead a Fraser Nash turret with .303in mgs and all fifty became Lancaster B.1s, by which time most 'ordinary' B.1s were also receiving the same 1,610hp Merlin 24s fitted to the B.7 in lieu of the less powerful Merlins previously installed.

Where are the missing Marks?
A few examples of the **B.1 (Special)** remained available to Bomber Command in the immediate post-war period for trials with 'Grand Slam' and 'Tallboy' deep penetration bombs. Apart from PD137 (which survived into 1952) most if not all of the others had been scrapped by mid-1948.

Most surviving examples of the Bristol Hercules-powered **B.II** had been removed from Bomber Command's Order of Battle by mid-1944 but continued to serve with training units until war's end. Only a handful lingered on into the post-war era and it would appear that the last two flying examples were the test beds LL735 and LL736 which were finally struck off charge (SOC) in March and June 1950 respectively.

Both the Lancaster **B.IV** and **B.V** were much-modified Marks – so much so that they were renamed to become the Lincoln B.1 and B.2 respectively and as such fall outside of this book's parameters.

Nine B.IIIs were converted to Lancaster **B.VI** standard which were used to trial higher-powered Merlin engines. A few survived as test beds into the post-war period.

The designations Mark **VIII** and **IX** were held in reserve for further Lancaster development if required.

Bomber Command Lancasters

This photo illustrates the familiar colour scheme in which the vast majority of wartime-built Lancasters emerged from the factories, i.e. the Temperate Land Scheme of Dark Earth and Dark Green upper surfaces with Night (black) under surfaces and fuselage sides to Pattern No.2. All that was then required was its unit code. Seen nearing completion at Vickers-Armstrong's (Chester) facility in early 1945, B.I, PA280, has yet to have its serial number applied, although it has been chalked onto the wing leading edge and forward fuselage. PA280 was received by No.227 Squadron in March 1945 and SOC in June 1947.
Newark Air Museum

At the close of 1945 the RAF's Lancaster force consisted of Marks B.1, B.1(FE), B.3, B.7 and B.7(FE), while the seven Canadian-built and operated B.10 squadrons, previously with No.6 Group (RCAF), had redeployed to Canada from June 1945 in preparation for 'Tiger Force'. (Other Canadian units equipped with the B.1 and B.3 still remained in Britain and were transferred to No.1 Group prior to No.6 Groups's disbandment on 31 August 1945. They and numerous other Allied bomber units having been heavily involved in ferrying troops and ex-POWs home from Europe.)

By the end of 1945 Bomber Command had slimmed down drastically having lost Nos.4, 5, 6 and 8 Groups although operationally its structure and function remained much as before albeit with its heavy bombers confined to just Nos.1 and 3 Groups. When the Command's peacetime needs were assessed in combination with those of the Middle and Far East it was considered that an operational total of approximately 445 Lancasters would be required. This figure also allowed for the replacement of American-supplied four-engined Liberator bombers which the RAF then operated in those regions and which would soon have to be returned to the US or paid for – which was not an option. Given the tendency of liquid-cooled Merlins to quickly overheat in tropical climates it was considered that FE-standard Lancasters, with tropicalized Merlins better able to cope with the heat, would be best suited to those regions while standard Lancasters would continue at home.

However, the figure of 445 proved to be optimistic and was soon reduced to 'about 270', presumably referring to a front-line strength that included immediate reserves. Given that Bomber Command would reduce to about twenty-two heavy bomber squadrons during 1946, with four others based in the Middle East, it would leave each with an average complement of ten aircraft and in fact many squadrons operated with an establishment of just eight aircraft, equivalent to a wartime Flight.

Until the Lincoln B.2 entered service in quantity the Lancaster B.1 became the primary workhorse of Bomber Command supplemented by the B.7. Together they steadily replaced the Packard Merlin-powered B.3, a measure taken to ensure that their Lancasters were powered by Rolls-Royce Merlins as an insurance against any disruption of supply from the USA. Additionally, given the surplus of available Lancasters, it was decided that Bomber Command's fleet should consist

of aircraft which had accumulated less than 150 hours flying time each. Consequently, from late 1945, quantities of 'young' yet surplus black and white-painted B.1(FE)s and B.7(FE)s, mostly fitted with dorsal turrets (which would have been removed for operations with Tiger Force had the war against Japan continued), began arriving with the squadrons to ultimately equip most of the Command's remaining Lancaster units.

Bomber Command Order of Battle as of January 1946 and January 1947

The tables overleaf illustrate the transition towards a standardization of Lancaster Marks that was underway in January 1946 and would continue for some months to come. They also show that the anticipated arrival of the new Lincoln B.2 (the B.1 didn't enter operational service) was already underway.

Taken in late-May 1945, this unidentified B.1 (Special) from No.617 Squadron carries a 22,000 lb 'Grand Slam' bomb and illustrates one of the few wartime Lancaster camouflage variations whereby its Temperate Land Scheme upper surface colours have been extended down the fuselage sides to meet with the Medium Sea Grey under surfaces to Pattern No.1. The upper halves of the port fin and rudder have been painted white. *Newark Air Museum*

A survivor of the Dam's Raid in May 1943, Lancaster B.1 (Special), ED906 'YF-A' is seen while serving with Scampton's Station Flight in 1945/46 – with its mounting arms for the 'Upkeep' mine (bouncing bomb) still in place and with post-war underwing serials applied. ED906 was briefly transferred to No.61 Squadron in 1946. It was finally SOC on 29 July 1947. *Newark Air Museum*

Lancaster B.1, LM274 'QR-F', of No.61 Squadron. Delivered in August 1944, this aircraft had completed sixty-nine operations by the war's end and was SOC in April 1946 having accumulated 690 hours flying time – hence it was barred from further service with Bomber Command given their policy of only accepting Lancasters with less than 150 hours flying time. When photographed in early 1946, LM274 retained its wartime camouflage and markings and had gained post-war underwing serials and a white panel on its port-inner engine donated, presumably, by a Lancaster FE. *Newark Air Museum*

■ BOMBER COMMAND ORDER OF BATTLE JANUARY 1946

No.1 GROUP

Binbrook, Lincolnshire
12 Sqn	Lancaster B.1, B.3	code PH
101 Sqn	Lancaster B.1, B.3	code SR

Coningsby, Lincolnshire
83 Sqn	Lancaster B.1, B.3	code OL
97 Sqn	Lancaster B.1, B.3	code OF

Faldingworth, Lincolnshire
300 Sqn	Lancaster B.1, B.3	code BH	Polish unit, disbanded 11/10/46

Metheringham, Lincolnshire
106 Sqn	Lancaster B.1, B.3	code ZN	Disbanded 18/2/46

Scampton, Lincolnshire
57 Sqn	Lancaster B.1, B.3	code DX	Lincoln B.2s arriving
100 Sqn	Lancaster B.1, B.3	code HW	

Sturgate, Lincolnshire
50 Sqn	Lancaster B.1, B.3	code VN
61 Sqn	Lancaster B.1, B.3	code QR

Waddington, Lincolnshire
9 Sqn	Lancaster B.7(FE)	code WS	Both units deployed to Salbani, India, 1/1946 to 5/1946)
617 Sqn	Lancaster B.7(FE)	code KC	

Leeming, Yorkshire
427 (Lion) Sqn	Lancaster B.1, B.3	code ZL	RCAF squadrons – both disbanded on 31/05/46
429 (Bison) Sqn	Lancaster B.1, B.3	code AL	

No.3 GROUP

Gravely, Cambridgeshire
35 Sqn	Lancaster B.1, B.3	code TL
115 Sqn	Lancaster B.1, B.3	code KO

Mildenhall, Suffolk
15 Sqn	Lancaster B.1, B.3	code LS	A few B.1(Specials) on charge for 22,000 lb bomb trials
44 Sqn	Lancaster B.1, B.3	code KM	Lincoln B.2s arriving

Methwold, Norfolk
149 Sqn	Lancaster B.1, B.3	code OJ
207 Sqn	Lancaster B.1, B.3	code EM

Tuddenham, Suffolk
90 Sqn	Lancaster B.1, B.3	code WP	
138 Sqn	Lancaster B.1, B.3	code AC	Code changed to NF in 12/1946

Mepal, Cambridgeshire
7 Sqn	Lancaster B.1, B.3	code MG
49 Sqn	Lancaster B.1, B.3	code EA

Although not part of Bomber Command, in 1946 No.205 (Heavy Bomber) Group operated Lancasters with the following squadrons in defence of the Canal Zone. The Group was disbanded on 31 March 1947.

- 40 Sqn   Lancaster B.7(FE)s received 1/46. Squadron disbanded 1/4/47.
- 104 Sqn   Lancaster B.7(FE)s received late 1945. Disbanded 1/4/47.
- 178 Sqn   Liberators replaced by Lancaster B.3s in late 1945. Squadron renumbered as No.70 Sqn in April 1946 receiving a few B.1/B.7(FE)s. Squadron disbanded 1/4/47.
- 214 Sqn   Lancasters received in late 1945, the squadron becoming No.37 Sqn on 15/4/1946. (No.214 Sqn reformed in the UK during 11/46). No.37 Sqn operated Lancaster B.3s until it received B.7(FE)s from mid-1946. No.37 Sqn disbanded on 1/4/1947 to re-emerge in 9/47 as a maritime squadron equipped with Lancaster GR.3s.

■ BOMBER COMMAND ORDER OF BATTLE JANUARY 1947

No.1 GROUP

Binbrook, Lincolnshire
9 Sqn	Lincoln B.2	code	WS
12 Sqn	Lincoln B.2	code	PH
101 Sqn	Lincoln B.2	code	SR
617 Sqn	Lincoln B.2	code	KC

Hemswell, Lincolnshire
83 Sqn	Lincoln B.2	code	OL
97 Sqn	Lincoln B.2	code	OF
100 Sqn	Lincoln B.2	code	HW

Waddington, Lincolnshire
50 Sqn	Lincoln B.2	code	VN
57 Sqn	Lincoln B.2	code	DX
61 Sqn	Lincoln B.2	code	QR

No.3 GROUP

Stradishall, Suffolk
35 Sqn	Lancaster B.1(FE)	code TL
115 Sqn	Lancaster B.1(FE)	code KO
149 Sqn	Lancaster B.1(FE)	code OJ
207 Sqn	Lancaster B.1(FE)	code EM

Upwood, Huntingdonshire
7 Sqn	Lancaster B.1(FE)	code MG
49 Sqn	Lancaster B.1(FE)	code EA
148 Sqn	Lancaster B.1(FE)	code AU
214 Sqn	Lancaster B.1(FE)	code QN

Wyton, Huntingdonshire
15 Sqn	Lancaster B.1(FE)	code LS	
44 Sqn	Lancaster B.1(FE)	code KM	Lincoln B.2 arriving
90 Sqn	Lancaster B.1(FE)	code WP	
138 Sqn	Lancaster B.1(FE)	code NF	

During the years 1948 and 1949 the ongoing replacement of Lancasters continued steadily until, by 1 January 1950, very few remained – all in No.3 Group – with Nos.7, 49, 148 and 214 Squadrons; by late March 1950 they too had gone. As an RAF bomber the Lancaster was now a memory.

However, Bomber Command continued to operate a handful of photo-reconnaissance Lancaster PR.1s that had equipped No.82 Squadron since October 1946 and was tasked with the aerial survey of Kenya and western Africa. Eventually, in October 1952, their task complete, the unit returned to the UK and by late 1953 its PR.1s had been replaced by Canberra PR.3s. In addition, No.683 Squadron also received the PR.1 which they operated in the Middle East from November 1950 until disbanded in November 1953. Reportedly, the last Lancaster in Bomber Command was PR.1, PA427, late of No.82 Squadron.

Lancaster B.1(FE)s of No.35 (Madras Presidency) Squadron. In early 1946 the notion of a goodwill tour of the USA by a Bomber Command squadron was considered and then developed to become Operation *Goodwill*. Number 35 Squadron was chosen to undertake the six-week tour using 16 Lancaster B.1(FE)s, their numbers apparently supplemented by employing Lancasters from other units. The squadron, supported by an accompanying Avro York, departed from Gravely in Cambridgeshire on 9 July 1946. The two images seen here were taken on 1 August 1946 during the US Army Air Force Day celebrations at Long Beach Field, Los Angeles. Lancaster 'H' is TW878 which later served with No.214 Squadron in 1949. It was sold for scrap in June 1950. *Newark Air Museum*

Top: Lancaster B.1(FE), TW659 'TL-M' of No.35 Squadron seen whilst participating in the tour of the USA (Operation *Goodwill*). Visible below the pilot is the Squadron Motif which bears the head of a winged horse, recalling the unit's co-operation with the cavalry in the First World War and which appeared on both sides of the cockpit. TW659 later served with No.7 and 214 Squadrons and was sold for scrap in August 1950. *via Tony O'Toole*

Above: Lancaster B.1(FE), TW887 'A3-T', of No.230 OCU which was sold for scrap in 1950. *Roger Lindsay*

Below: Two Lancaster B.1(FE)s with all turrets armed and manned – the date is unknown. TW907 'EA-L' only ever served with No.49 Squadron and was scrapped in January 1951. The other serial is not visible but 'QN' belongs to No.214 Squadron. *Newark Air Museum*

Two Lancaster B.1(FE)s, again with all turrets armed and manned. 'QN-A' is TW873 of No.214 Squadron accompanied by an unidentified aircraft from No.148 Squadron. Although the image is undated it is known that TW873 was allocated to this unit in October 1946 and that it made a forced landing on 23 April 1947 at Aylsham, Norfolk, with both port engines feathered. There were no fatalities but the aircraft was written off. *Newark Air Museum*

Lancaster PR.1, TW905 'G' of No.82 Squadron. This variant of the Lancaster was a photo-reconnaissance platform in which all three turrets were removed and faired over while much of the cockpit glazing was painted over, or replaced by metal panels, to help protect the crew from the searing heat of the equatorial sun. Painted in the black and white Tiger Force scheme, TW905 retains post-May 1942 wartime national markings, albeit the colour reproduction has been affected by the type of film used. It was sold for scrap in June 1950. *Tony O'Toole*

Two unidentified No.82 Squadron Lancaster PR.1s in flight over Kenya. Both are finished in painted Aluminium with post-war national markings applied. It appears that some latitude existed in deciding where individual identification letters should be placed. As a point of interest, the amount of cockpit glazing on these two examples is significantly less than on TW905.
Newark Air Museum

Coastal Command Lancasters

Above: Lancaster GR.3, RE168, No.203 Sqn – date and location not supplied. Converted to ASR.3 and later to GR.3 standard, this Lancaster served with Nos.179, 224 ('XB-M') and 203 Squadrons (CJ-F). It was sold for scrap on 29 December 1953.
Newark Air Museum

Opposite, top:
Lancaster ASR.3, RF310 'RL-A', of No.279 Squadron seen dropping a lifeboat on 10 December 1945. After the unit disbanded in March 1946, its code 'RL' was later adopted by No.38 Squadron in Malta from July 1946 to 1951. Other images exist to show that RF310 still retained front and rear guns at this time despite a requirement to remove them from the Mark to provide a less impeded view for the crew. Subsequently used by No.1438 Flight at Pegu, Burma, RF310 was destroyed there when its undercarriage collapsed while taking off on 4 March 1946.
Newark Air Museum

As previously mentioned the end of the Second World War meant that aircraft supplied to Britain by the USA under the terms of the Lend-Lease Act were to be returned to the US, scrapped, or purchased outright: the latter, readers will recall, was not an option for Britain. The impending return of Lend-Lease Liberators, Hudsons, Catalinas and Fortresses would inevitably leave RAF Coastal Command with a maritime patrol and air-sea-rescue (ASR) capability gap which their existing, but relatively few, Sunderland flying boats and twin-engined Vickers Warwick ASR aircraft alone would not be able to fulfill. Therefore, given the large numbers of ex-Bomber Command Lancaster B.IIIs that quickly became surplus to requirements in 1945, it was decided that a number should be converted as soon as possible to the ASR role, followed in due course by others to be converted to the maritime reconnaissance role – or general reconnaissance (GR) as it was then termed.

The origin of the ASR Lancaster actually commenced with preparations for 'Tiger Force' for which an adequate ASR capability had to be found to assist downed aircrews as they traversed vast tracts of the Pacific Ocean in the war against Japan, and for which the existing ASR Warwick was inadequate. The Packard Merlin-powered Lancasters selected for conversion were to include the ability to carry a Cunliffe-Owen 30ft 6in lifeboat which, when released, would descend by parachute. Tested by Avro using a Lancaster B.1, the lifeboat was fitted beneath its (closed) bomb bay doors and secured to a heavy-duty beam in the bomb bay that was originally intended to carry a 4,000 lb bomb. A strut then descended from the bomb bay, through small flaps cut into the bomb bay doors, and attached to the boat. Following these tests, Cunliffe-Owen Aircraft Ltd was contracted to convert 130 Lancaster B.IIIs to ASR.III standard with the first arriving at their Eastleigh Airport facility, near Southampton, in early 1945.

While the B.III's existing H2S Mk.III radar was considered adequate for purpose and retained (despite being prone to the 'clutter' created by various sea states), several other alterations were also made including the fitting of two sets of directional antennas with which to pick-up distress messages and triangulate their source. The mid-upper turret was removed, as too were the guns from the remaining turrets which improved observation from those positions considerably. To further improve observation, additional windows were provided in the port fuselage forward of the tailplane, while on the opposite side a window was added to the fuselage door.

Two units were due to receive the Lancaster ASR.3 during 1945, Nos.179 and 279 Squadrons in readiness for 'Tiger Force', but the ASR.3 was slow to enter service and by September 1945 only a single flight of No.279 Squadron had been equipped with the type. When the war in the Pacific ended (and the need for 'Tiger Force' evaporated) the rate of conversion slowed further, consequently No.179 Squadron didn't receive ASR.3s until February 1946, some of which were transferred from No.279 Squadron when it disbanded a month later.

With ASR.3 conversions underway, in 1946, Coastal Command had to address the

An undated image of uncoded Lancaster, RF289. Converted to an ASR.3 in 1947 and to a GR.3 in 1948, RF289 is seen after conversion with late-war roundels applied in six positions. RF289 went on to serve with No.120 Sqn from February 1949 and also the School of Maritime Reconnaissance (SMR) from May 1951 and was scrapped on 14 July 1955.
Tony O'Toole collection

Lancaster GR.3 SW283, 'OZ-U' of No.210 Squadron. Converted to an ASR.3 and first operated by No.279 Squadron, it later went to No.179 Squadron. When No.210 Squadron reformed in June 1946 it absorbed aircraft from No.179 and kept the latter's unit code 'OZ' which was retained into1951. No.210 was the first squadron to operate the GR.3, but at what point SW283 was upgraded isn't recorded.
Tony O'Toole collection

Another photo of GR.3, SW283, now 'L-U' seen 'out to grass at Langar' (as written on the back of the photo) in June 1952. Late of No.210 Squadron, their code had changed from 'OZ' to 'L' in 1951. SW283 still had life to spare and went on to serve with the SMR at St Mawgan coded 'H-Q'. It was sold for scrap in May 1957.
Author's collection

Lancaster GR.3, SW376 'B-A' of No.203 Squadron *circa* 1952. Converted to an ASR.3 and then to GR.3, this unit's code had been 'CJ' until 1951 when it changed to 'B'. SW376 subsequently went to the SMR coded 'H-Y'. In common with several GR.3/MR.3 Lancasters still nominally extent, SW376 was finally SOC in May 1957, at Wroughton, where many lingered until finally disposed of. *Tony O'Toole collection*

Meanwhile in Malta…
A clipped 'Box' Brownie image of Lancaster ASR.3, RF300 'RL-K' seen at Luqa, Malta in April 1949 with Medium Sea Grey and white camouflage and still with wartime roundels in six positions and No.38 Squadron codes, although it was also used by No.37 Squadron too. Returned to the UK in May 1949 for conversion to GR.3 standard, a later examination found RF300 to be unsuitable for conversion and so was scrapped in September 1950. *Newark Air Museum*

Nose of Lancaster ASR.3, RF269 'RL-W', of No.38 Squadron seen at Luqa in May 1949 displaying its Squadron motif featuring a heron in flight. Returned to the UK on 27 June 1949, RF269 was subsequently converted to GR.3 standard and issued to the SMR coded 'H-L'. It was SOC on 17 July 1956. In the background sits ASR.3, RE123 'RL-R', wearing an earlier camouflage scheme; not further modified, RE123 was scrapped in August or September 1951. *Newark Air Museum*

problem created by the pending loss of GR Liberators during 1946 and 1947, a role for which the basic ASR.3 was unsuitable. Ultimately the problem would be answered by Avro's 'maritime Lincoln', a much-modified derivative of the Lincoln B.2, so modified in fact that it justified, and received, a new name: the Avro Shackleton. It was of course recognised that several years would elapse before the Shackleton was ready for operational service, a time gap which in the interim would be filled by the Lancaster GR.3.

Virtually identical externally (camouflage apart) to the ASR.3, the GR.3 incorporated a higher-definition, purpose-designed air-to-surface-vessel radar, and optional auxiliary fuel tanks in the bomb bay which was itself reorganised to better accommodate various flares, weaponry and sonobouys (first used during the Second World War). While still fully capable of carrying an airborne lifeboat, the GR.3 did so much less frequently than the ASR.3 (although perversely a majority of the GR.3 photos originally collected for this book showed them with a lifeboat fitted – probably because most were taken at 'open days' or other public events.) Later the GR.3 was redesignated to become the Lancaster MR.3 (Maritime Reconnaissance). Whether this change related to a further upgrade in capability or not isn't clear, it might simply reflect a change that was gradually applied to maritime patrol aircraft – possibly brought about once the Shackleton GR.1 became the MR.1 shortly before it commenced service trials in April 1951. It isn't clear either as to how many Lancaster GR.3/MR.3 conversions were made, the situation being muddied by the fact that several ASR.3s were upgraded to GR.3 and in some instances to MR.3 standard too.

By 1 July 1948, five operational maritime reconnaissance Lancaster squadrons were in existence deployed as shown in the bottom table on the opposite page. All were primarily GR.3 equipped although in 1948 many ASR.3s remained too. The GR.3 continued with all five squadrons pending their steady replacement by more modern types which, starting with No.120 Squadron, began in 1951. The last RAF squadron to deploy operational Lancasters of any sort was No.38 whose last examples survived into early 1954.

COASTAL COMMAND UNITS 1945-1956

Unit	Mk	Service	Notes
18 Sqn	ASR.3	1/9/46 – 15/9/46	Unit reformed in Palestine by renumbering No.621 Sqn. Disbanded two weeks later but remained in-situ and absorbed by No.38 Squadron.
37 Sqn	GR.3	9/47 – 9/53	Reformed in Palestine with GR.3s for MR duties and patrol of eastern Mediterranean. Redeployed to Luqa, Malta from April 1948. Shackleton MR.2 from Aug 53.
38 Sqn	ASR.3 GR.3	7/46 – 1949 7/46 – 2/54	Based at Luqa, this unit received about 12 ASR.3s, some late of 279 Sqn and No.1348 Flight. Also rec'd GR.3s, the second squadron to use the Mark. No.38 Sqn sent detachment to Palestine on 15/9/46 to join ex-No.18 Sqn. Returned to Luqa April 1948. No.38, the last RAF Lancaster squadron, received Shackleton MR.2s from 9/53.
120 Sqn	ASR.3 GR.3	11/46 –?? 1947 – 4/51	Reformed 1/10/46 from No.160 Sqn. Initially operated mixed unit of GR Liberators and Lancaster ASR.3s. Liberators gone by mid-1947. Palestine 11/47. Returned to Leuchars. Kinloss by 12/50 prior to receiving Shackleton MR.1s in early 1951.
160 Sqn	ASR.3	9/46 –10/46	Based at Leuchars. ASR.3s supplemented GR Liberators in 8/46. 160 Sqn disbanded to become No.120 Sqn 10/46.
179 Sqn	ASR.3	2/46 – 9/46	Warwick-equipped at St. Eval, this unit received a few ASR.3s from No.279 Sqn, the Lancaster element becoming No.179X Sqn. In 5/46 the Warwicks were disposed of and the unit reverted No.179 Sqn until disbanded on 30/9/46.
203 Sqn	ASR.3 GR.3	8/46 – ?? 8/46 – 3/53	Based at Leuchars, this unit received Lancaster ASR.3s and GR.3s from 8/1946 and had replaced Liberators by 10/46. In 1/47 the unit relocated to St. Eval, moving to Topcliffe in 8/52. GR.3s replaced by Neptune MR.1 by 3/53.
210 Sqn	ASR.3 GR.3	6/46 – 1949 6/46 – 12/52	Reformed 1/6/46 with six ASR.3s based at St. Eval. Received ASR.3s from No.179 Sqn in 9/46. No.210 also obtained GR.3s in 6/46, becoming the first squadron to use the Mark. Replaced by Neptune MR.1 in early 1953.
224 Sqn	ASR.3? GR.3	1946 – ?? 10/46 -11/47	Based at St. Eval, GR Liberators replaced by Lancaster GR.3s in 11/46. Disbanded 10/11/47.
279 Sqn plus 1348 Flt	ASR.3 ASR.3	9/45 – 3/46 – 5/46	In 9/45 No.279 Sqn was at Beccles, Suffolk, where it received early Lancaster ASR.3s. In 12/45 a detachment was sent to Burma to provide ASR cover for the area. While there the parent unit disbanded and the detachment was absorbed by No.1348 (ASR) Flight which disbanded on 15/5/46. Surviving ASR.3s ultimately went to No.38 Sqn in Malta.
621 Sqn	ASR.3	4/46 – 9/46	Based at Aqir, Palestine, No.621 Sqn received ASR.3s in late-April 1946. August. Renumbered on 1/9/46 the unit became No.18 Sqn.
ASWDU	B.1, B.3, ASR.3, GR.3/MR.3	Formed 1/45 Disbanded 4/70	The Air-Sea Warfare Development Unit (ASWDU) moved to Ballykelly on 26/5/48 then to St. Mawgan in 5/51
JASS Flight	ASR.3, GR.3	Formed 11/45 Disbanded 6/71	The Joint Anti-Submarine School Flight (JASS) formed on 19/11/45 at Ballykelly to practice anti-submarine tactics.
236 OCU	B.1, ASR.3, GR.3/MR.3	Formed 7/47 Disbanded 9/56	Formed 31/7/47 at Kinloss to train GR aircrews. On 1/10/56 it combined with No.1 Maritime Reconnaissance School (No.1 MRS) to form the Maritime Operational Training Unit (MOTU).
SMR/ 1 MRS	GR.3/ MR.3	Formed 5/51 Disbanded 9/56	The School of Maritime Reconnaissance / No.1 Maritime Reconnaissance School was formed on 1/5/51 at St. Mawgan to provide all aspects of air-sea warfare. On 1/10/56 it combined with No.236 OCU to form the MOTU
MOTU	MR.3	Formed 10/56 Disbanded 7/70	The Maritime Operational Training Unit formed on 1/10/56 (by which time few if any Lancasters remained except RF325 which was disposed of before the end of the month).

OPERATIONAL RAF MARITIME RECONNAISSANCE SQUADRONS 1 JULY 1948

Squadron	Base	Command/Air Force	Squadron	Base	Command/Air Force
120 Sqn	Leuchars	18 Group, Coastal Command	37 Sqn	Luqa	Middle East Air Force
203 Sqn	St. Eval	19 Group, Coastal Command	38 Sqn	Luqa	Middle East Air Force
210 Sqn	St. Eval	19 Group, Coastal Command			

Above: Lancaster GR.3, RE167 'F', No.37 Squadron, Malta, 1950 – although this unit was allocated the code 'LF' it was not used after October 1946 at which time it was still a bomber squadron. It would seem that once it became a maritime unit No.37 Squadron only ever used a single-letter code with only the squadron badge (encompassing their hooded hawk motif) helping to identify the unit. RE167 joined the unit in 1949 and was SOC in June 1953. *Roger Lindsay*

Below: Lancaster GR.3, RE185 'D', seen in Malta in 1949/1950. This Lancaster was allocated to No.37 Squadron in 1949 and then to No.38 Squadron at a later date, which poses the question as to which unit was operating this aircraft when the camera clicked. Subsequently returned to the UK, RE185 presumably went into storage before being acquired and scrapped by Enfield Rolling Mills in September 1956. *Roger Lindsay*

Opposite page: Two images of Lancaster GR.3, RF313 'Y' of No.38 Squadron seen in February 1953. This Lancaster had been converted to an ASR.3 in 1945 and allocated to No.279 Squadron in December 1945. It was with No.38 Squadron by December 1947. Later converted to GR.3 standard, RF313 served with Flight Refuelling Ltd in 1951/52 prior to being returned to No.38 Squadron, following which it went to the School of Maritime Reconnaissance (SMR) coded 'H-N'. It was sold for scrap on 22 May 1957. *Both Newark Air Museum*

Lancaster GR.3, TX273 'U' from No.38 Squadron *circa* 1950/51. TX273 was the last serial number to be allocated to a Lancaster – which of course is **not** to say that this was the last Lancaster completed, an entirely different question with the latter distinction belonging to a Lancaster B.1 in the TW647-TW911 serial range, the last of which were delivered in March 1946. TX273 was scrapped in September 1956. *Newark Air Museum*

Lancasters with the School of Maritime Reconnaissance (SMR).

Formed originally as the SMR in May 1951 at St. Mawgan, its name was later changed (on an, as yet, undiscovered date) to No.1 Maritime Reconnaissance School (MRS). To most, however, it simply remained as the SMR until 1 October 1956 when it was combined with No.236 OCU to become the Maritime Operational Training Unit (MOTU). Whether SMR or MRS the unit code 'H', accompanied by a single identifying letter, was retained.

Lancaster GR.3, RE164 'H-U', seen at Odiham on 15 July 1953 on the occasion of the RAF Coronation Review. One of four SMR/MRS Lancasters present in the static display that day, the others were RE181, RF325 and SW334; the latter, coded 'H-H', can be seen just beyond RE164. Having previously served with Nos.279, 179 and 224 Squadrons, RE164 went on to serve with the ASWDU, JASS and SMR before going to Flight Refuelling Ltd. It was SOC on 24 August 1956.
Tony O'Toole collection

Having appeared at the Coronation Review in 1953, ex-GR. (now) MR.3, SW334 'H-H' is seen close to St. Mawgan in far less happy circumstances on 21 May 1955 following an unsuccessful three-engined overshoot – fortunately without fatalities; this proved to be the penultimate loss of an RAF Lancaster. MR.3, RE211 became the last when its pilot was forced to abort his take-off run at St. Mawgan on 2 February 1956 by raising the undercarriage to stop after the aircraft began skidding on snow and ice. *Newark Air Museum*

An undated photo of six SMR Lancasters. *Newark Air Museum*

Left: Previously operated by No.236 OCU coded 'K7-C', Lancaster GR.3, RE186 'H-C' is seen following its transfer to the SMR. This photo was taken *circa* 1953/54 at Blackbushe, Hampshire. *Tony O'Toole collection*

Above: On the same date RE186 is shown from a less usual angle clearly demonstrating its stained, workaday appearance. The projection below the tail turret was a feature of most maritime Lancasters and contained a strike camera and flare chute, the flares being for night photography. RE186 was sold for scrap on 22 May 1957. *Tony O'Toole collection*

Below: Lancaster MR.3, SW366 'H-Z' seen at Blackbushe in May 1956 wearing the final maritime colour scheme to be applied to RAF Lancasters. Introduced in 1955, the overall Dark Sea Grey Scheme was accompanied, in this case, by white serials and codes although at least three MR.3s received red underwing serials and codes (outlined in white) with plain red serials on the fuselage. *Roger Lindsay*

Right: SW366 seen in the same year at St. Mawgan. By now the Lancasters' time in the RAF was running out. SW366 was sold for scrap on 22 May 1957. *Tony O'Toole collection*

RF325: The RAF's last Lancaster

Many images survive of RF325 with those shown here illustrating it as an ASR.3, GR.3 and MR.3. These photographs are unusual in that they tend to show that this Lancaster, visually at least, unlike so many others, seemed to improve with age and so it remains a matter of regret that when RF325 was finally retired in October 1956 there was no thought of preserving it for posterity. RF325 made its last flight, to Wroughton, in the middle of October and was SOC on 11 July 1957.

Top: ASR.3, RF325 'P9-J', with the ASWDU in 1949 in the Temperate Sea Scheme with wartime national markings. *Tony O'Toole collection*

Centre: GR.3, RF325 'H-D', re-finished in the Coastal Command grey and white scheme with post-war national markings while serving with the SMR during major NATO exercise, Exercise *Mariner*, in 1953 for which tactical markings were applied to the fuselage of many participating aircraft. It is believed that these were yellow and red with the yellow band leading. *Author's collection*

Bottom and opposite page, top: RF325, by now redesignated as an MR.3 and repainted in the overall Dark Sea Grey scheme and still coded 'H-D'. Seen in mid-October 1956 at the SMR, these two photos were taken just an hour or so before RF325 was rolled out of its hangar at St. Mawgan for the last time prior to its final flight. *Both Newark Air Museum*

Opposite, centre: RF325 seen later the same day en route to Wroughton for scrapping. *Roger Lindsay*

Opposite, bottom: Wroughton, 1957. It is thought that RF325 is the third Lancaster from the rear of the line. Other identifiable Lancaster serials are: SW367, SW368 and SW376, while in the next field several Lincolns and Mosquitoes are visible. *Author's collection*

Second-line Lancs Miscellany

Lancaster B.7(FE), NX668. Seen at Langar, Nottinghamshire, in June 1952. Problems arise regarding the fate of this aircraft which hitherto had not been allocated to any unit other than MUs. Sources state that NX668 became WU-24 with the *Aéronavale* in 1952, however, the same sources insist that B.1(FE) PA389 became WU-24. Perhaps NX668 was found to be unsuitable for conversion and subsequently replaced by PA339! If so, doubtless the B.7 would have been SOC soon afterwards.
Newark Air Museum

Lancaster B.1, PD328 *Aries* from the Empire Air Navigation School (EANS) – it would later receive the code 'FGFA' ('FGF' being the unit's code while 'A' was *Aries*'s identifying letter). Occasionally referred to as a Lancastrian due to the nose and tail fairings fitted in the spring of 1945, it remained, nevertheless, a Lancaster. Given that PD328 displays unshrouded exhausts, lacks underwing serials and roundels as well as an eight-pointed star near to the tip of its nose, it is likely that this image was taken shortly after it was modified at Waddington. Of interest perhaps is the fact that immediately following PD328's modifications, for a (very) brief period it was unnamed, retained shrouded exhausts and carried Pacific-style national markings – a legacy of its pioneering round-the-world flight which had commenced in October 1944 and encompassed the Pacific. *Aries* was scrapped either in January 1947 or August 1948 according to the source consulted.
Tony O'Toole

Despite the initial post-war need for quantities of Lancasters by both Bomber Command and Coastal Command, several hundred others still remained available many of which would soon find employment with training, trials or test units. Even then many more remained unallocated in outside storage where they languished and often deteriorated to a point whereby, externally at least, they seemed fit only for scrap; yet dozens of these would be reconditioned, virtually rebuilt, and supplied to France years later.

Although it is not feasible in a book of this size to list every British unit that fielded a Lancaster during the post-war era, the following provides an indication of its widespread use in non-operational second-line duties.

They include the: Aeroplane & Armament Experimental Establishment, Air-Sea Warfare Development Unit, Aircraft Torpedo Development Unit, Blind Landing Experimental Unit, Bomber Command Film Unit, Bomber Command Instructors School, Bombing Development Unit, Bombing Trials Unit, Central Bomber Establishment, Central Flying School, Central Gunnery School, Central Navigation and Control School, Central Signals Establishment, Empire Air Armament School, Empire Air Navigation School, Empire Central Flying School, Empire Flying School, Empire Test Pilots School, No.780 Squadron Fleet Air Arm (RN), No.1323 (Automatic Gun Laying Turret Training) Flight, Joint Anti-Submarine School Flight, No.230 Operational Conversion Unit, No.231 OCU, No.236 OCU, No.6 Operational Training Unit, Radio Warfare Establishment, RAF Flying College, Royal Aircraft Establishment.

Lancaster B.7(FE), NX697 'FGG-G', Central Navigation School (CNS), seen at Middleton St George in September 1949. The CNS was originally formed in 1942 but, in late October 1944, it was renamed and became the EANS tasked with scientifically investigating the advancement of navigational instruction and to explore the development of new techniques intended to overcome the problems associated with worldwide navigation. The EANS was renamed to become the CNS again on 31 July 1949, however, in February 1950 it changed once more becoming the Central Navigation and Control School (CN&CS). NX697 was sold for scrap in August 1954. *Tony O'Toole*

Lancaster B.7(FE), NX773 'FGG-C' *Capella* from the CN&CS at Istres, France, on 8 June 1951. NX773 had served with No.9 Squadron from November 1945 before being sent for storage at various maintenance units. Subsequently resurrected, NX773 arrived with the CN&CS in February 1951 painted in the post-war Medium Sea Grey and black Bomber Scheme, its training function highlighted by the yellow bands painted on the rear fuselage and around both wings between each pair of engines. Additionally, albeit hard to discern, the mid-upper gun turret has been replaced by an additional astrodome. *Author's collection*

A closer view of NX773's nose on the same date (*Capella* is the name of the third brightest star in the *northern* celestial hemisphere). Just visible to the extreme left of this image is a portion of the starboard yellow training band where it wraps around the leading edge of the wing. *Capella* was sold for scrap in June 1954. *Author's collection*

An undated image of CN&CS Lancaster B.7(FE), NX648 'FGG-O', in Malta. Having served originally in the Middle East, NX648 was subsequently returned to the UK and stored prior to being allocated to this unit. Just visible is the astrodome in lieu of the mid-upper gun turret, while a small portion of the yellow fuselage band can be seen aft of the starboard rudder. *Author's collection*

Lancaster B.7(FE), NX737 'FDI-B' from the Central Flying School (CFS) seen visiting Cranwell in June 1947 having already acquired the (then) very new post-war grey and black Bomber Scheme introduced barely two months earlier. NX737 previously served in the Middle East with Nos.40 and 70 Squadrons prior to joining the CFS in June 1947. It became maintenance airframe 6736M in April 1950. *Newark Air Museum*

Above: A post-March 1947 image of Lancaster B.1(FE), PA380, serving with the Central Signals Establishment (CSE) to which it had been issued following repairs caused during the severe winter gales of March 1947. At this time the CSE used the codes '4S' and 'V7' thus PA380 would have been coded either '4S-B' or 'V7-B'. It was scrapped in April 1950. *Newark Air Museum*

Below: Lancaster B.7 RT680: date, location and unit not known. This aircraft served first with the Empire Flying School (EFS) and then, from January 1949, with the Empire Air Armament School (EAAS), both of which were absorbed into the RAF Flying College (RAFFC) on 31 July 1949. RT680 displays yellow bands around the wings and rear fuselage and its turrets are armed. *Newark Air Museum*

Lancaster B.1, LL780 'DF-N', which served with the Central Bomber Establishment (CBE) during 1946/47 and was sold for scrap in June 1948. LL780 was used to investigate methods of improving bomber self-defence capabilities by the installation of gun-laying radar and two twin 20mm cannon-armed barbettes, the latter remotely controlled from a much-altered rear gun position. Almost obscured, the fuselage serial number, LL780/G, is just visible aft of the code 'N'. *Newark Air Museum*

Above: Lancaster B.1(FE), later PR.1, PA474 'M', seen at Cranfield, Bedfordshire, while serving with the College of Aeronautics. Having served with No.82 Squadron (whose badge, colours and code it retained into 1964), PA474 was loaned to Flight Refuelling Ltd in May 1952 for conversion to a pilotless drone. That conversion never occurred but the Lancaster did receive a major service before it was transferred to Cranfield, on 7 March 1954, where it was eventually employed to conduct research into boundary layer airflow using various wing sections affixed to the top of the fuselage. During September 1962, two gas turbine engines, used to suck boundary-layer air from the section under test (in this instance a Handley Page laminar flow wing section) were installed – their exhausts being visible just forward of, and below, the roundel. Replaced by Lincoln B.2 RF342, PA474 was flown to No.15 MU on 22 April 1964 where it was subsequently painted in wartime colours, the beginnings of a restoration process which would see PA474 become the world's most famous Lancaster which now resides with the BBMF. *Author's collection*

Below: Lancaster PA474 seen trialling another wing section while with the College of Aeronautics at Cranfield. The two gas turbine engine exhausts are absent. *Author's collection*

Above and below: To include these or not – that really was a question! One can almost hear the anguished sigh of veteran Lancaster enthusiasts as they see NX612 in print again. However, these images were part of a series of recognition photos intended to familiarise personnel with both a new Mark and a new colour scheme. The point, ultimately, is that both images are sharp, offer clear definition and reveal the stark simplicity of the FE colour scheme, ideal for the modeller, and thus included for that reason if no other! So, the details. Lancaster B.7(FE), NX612 '9X-', belonged to No.1689 (Ferry Pilot Training) Flight and was the second B.7 to fly. It was subsequently allocated to the Austin Motor Company Ltd and sold for scrap in February 1950. *Both Author's collection*

Lancaster B.7(FE), NX661. Seen stored at 20 MU Aston Down, NX661 awaits its fate which wasn't long in coming as it too fell victim to the severe winter gales of March 1947 and was SOC eight months later.
Newark Air Museum

Lancaster B.1(FE) TW655, seen at the *Daily Express* Exhibition of 19 to 21 June 1951, the theme of which was 'Fifty Years of Flying'. TW655, with turrets armed for the occasion, was in poor external condition having been flown straight into outside storage some four or five years earlier.
Newark Air Museum

TW655, appearances notwithstanding, went on to serve with the *Aéronavale* in 1952 as WU-17. It didn't long survive though as it was damaged while landing at Port Lyautey, Morocco and SOC in October 1953.
Newark Air Museum

In 1948 Western Union agreements led to the acquisition by France of fifty-four refurbished ex-RAF Lancaster B.1s and B.7s (becoming WU-01 to WU-54) with deliveries commencing in early 1952. Five further B.7s were to be supplied as FCL-01 to -05 which, while undergoing refurbishment in the UK were allocated the temporary 'B-class' marks G-11-69 to G-11-73 inclusive, with G-11-69 allotted to B.7, RT673. This photo, taken at Langar on 2 June 1952, is a mystery because G-11-69 isn't a B.7, the presence of the mid-upper turret with a fairing making this a B.1 (the B.3 can be discounted by virtue of its Packard-built Merlins). Sadly, it has not been possible to discover its true identity.
Author's collection

The Film

In April 1954 the shooting began of what would become probably the most famous and iconic of British post-war, war-related films: *The Dam Busters* – still regularly broadcast on network television with only a certain dog's name edited out to reflect modern sensitivities.

Four Lancaster B.7s, NX673, NX679, NX782 and RT686 were returned to duty and flown to Hemswell for use in the film, although NX782 would remain 'unmodified' so-to-speak and retained its original structural appearance in order to represent Guy Gibson's Lancaster during an earlier stage of the film prior to the Raid itself. The first three Lancasters listed had appeared in the film *Appointment in London*, released in 1953, for which they each received a dark green disruptive camouflage pattern applied to their existing overall Medium Sea Grey upper surfaces. The following images were taken during the shooting of the film in 1954. For reference 'AJ-G' was NX679 and 'AJ-P' was NX673. All four Lancasters were scrapped in July 1956.

All photographs Newark Air Museum

SECOND-LINE LANCS MISCELLANY 29

The Lancaster in RCAF Service

The Victory Aircraft Company at Malton, Ontario, completed 430 Packard Merlin-powered Lancasters during and immediately following the close of the war in Europe. All were B.10s, a Mark reserved for Canadian production with serial numbers within the groups KB700 to KB999 and FM100 to FM229 inclusive. Several were dispatched to Britain from where they operated against Germany during the last twelve months of the war. Following Hitler's demise the Allies were able to focus their attentions against the Japanese in the Pacific which included the creation of Tiger Force. With Japan's capitulation in August 1945, the need for Tiger Force disappeared and so it was officially disbanded at the end of October. For the Royal Canadian Air Force (RCAF) this meant the return of a number of Lancaster B.10s which were serving abroad and the retention of many others which still remained in Canada at the close of hostilities. As Canada returned to a peacetime footing the RCAF assessed its future roles and requirements and how to employ the assets it possessed, one of which was a surplus of Lancaster bombers, a role for which there seemed to be no immediate need and consequently most Lancasters were ferried into storage pending a decision on their future.

In total the *post-war* RCAF seems to have placed 230 Victory-built Lancasters (excluding Lancastrians) on its inventory, though sources contradict the precise amount by, plus or minus, one or two. Although a quantity of about 230 is correct a very large proportion never returned to flying duties at all with disposals commencing from early 1946 and continuing into 1947 and 1948. (Additionally a further seventy or so Canadian-built FM-serialed Lancasters remained in store at RAF maintenance units across England and Wales, most of which were struck off charge [SOC] during May and June 1947.) Other than those specifically marked for possible future use, few Lancasters survived in store beyond August 1950, although presumably those that were scrapped would at least provide a source of spares for those that remained.

Canada is the world's second largest country, eighty per cent of which consists of forest and tundra, and in the 1940s large tracts of the country's inhospitable far northern areas remained improperly mapped or not mapped at all; consequently its national borders in those regions remained ill-defined. The Canadian government understood that an urgent survey was required and that the most efficient method of conducting it was from the air, for which their most appropriate aircraft was a suit-

Lancaster B.10 FM153 arrived in England on 3 June 1945, by which time the war in Europe was over. Following a period of storage at No.218 MU, FM153 was flown back to Canada in August 1945 and put in storage until SOC on 19 January 1948. This is how early post-war B.10s looked originally, drab compared to the colourful RCAF examples that would appear in the near future. A Martin mid-upper turret can be seen just forward of the fuselage roundel.
Via Simon Watson

ably modified Lancaster. In 1946 two airframes were adapted to act as prototypes and they in turn were followed by the first of the Lancaster 10 sub-types to be modified for post-war RCAF service, the Mk.10P. Optimised for photo-reconnaissance duties, the Mk.10P went on to prove itself as an excellent aerial survey platform.

Canada's borders extend not only from east to west, from the Atlantic to the Pacific, they also extend far to the north of the arctic circle too, to the Arctic Ocean itself, on the other side of which lies the (then) Soviet Union. As post-war tensions between the West and the Soviet Union grew, so did concerns over possible Soviet incursions into Canada's polar regions. Consequently the Lancaster 10AR, a derivative of the Mk.10P, was developed from 1952 which was equipped with several cameras, surveillance equipment and an AN/APS-42 navigation and weather radar fitted beneath a specially extended nose, all of which allowed the Mk.10AR to conduct sophisticated reconnaissance flights over the ice-bound northern regions.

A further derivation of the Mk.10P was the Lancaster 10N which was optimised to train navigators, a particularly relevant skill when flying over vast areas of often featureless and inhospitable land, sea, or ice in an age when satellite-based navigational systems were science fiction.

In addition to the Mk.10P, Mk.10AR and Mk.10N, the Lancaster 10BR was also developed. The Mk.10BR (Bomber Reconnaissance) and its derivatives will always be associated with maritime reconnaissance and search-and-rescue (SAR) duties. However, as part of its original remit the 10BR retained a bombing function in addition to long-range reconnaissance and SAR roles. In reality though the need to conduct bombing operations soon become an irrelevance particularly once Canada's powerful neighbour, the USA, developed its Strategic Air Command (SAC) into a weapon of enormous offensive capability following the appointment of General LeMay as SAC's commanding officer in late 1948. The Mk.10BR was subsequently used as a basis from which to

Developed from the Mk.10P, three Lancasters were converted to Mk.10AR standard, all of which received a distinctive 40in nose extension plus navigational, electronic and photographic equipment; each had their turrets removed and their positions faired over. Seen head-on following conversion in 1952, KB882 displays the one-piece Perspex nose fairing fitted initially which would later be replaced by a Perspex and metal fairing. At this time KB882 mounted a thimble-shaped radome housing an AN/APS-42 radar under its nose, either side of which black Perspex triangular-shaped sections were added – the purpose of which remain unknown to the authors. At a later date they were removed and a lengthened radar fairing fitted. A second radome was added under the aft fuselage. *Newark Air Museum*

Having emerged as a Mk.10AR, KB882 was allocated to No.408(P) Photographic Squadron whose unit code 'MN' is displayed on the fuselage. This photo was taken at the same time as the preceding one and shows a covered camera port (one of three) aft of the bomb bay with another camera aft of the tail wheel, plus an array of radio antenna/aerials. The bomb bay was fitted with two long-range fuel tanks and a luggage/spares pannier. Retired in April 1964 and SOC in May, KB882 was flown into preservation on 14 July 1964. *Newark Air Museum*

Lancaster 10AR, KB976 'MN-976' displays the later Perspex and metal nose fairing, lengthened radar mounting beneath the nose, curved rear-fuselage observation windows (fitted to all three examples) and a second dome aft – associated with a UPD-501 passive intercept receiver. There were other additions too, both internal and external, though not all were necessarily fitted to the trio simultaneously. Presumably some items were added or removed according to need, such as the device beneath the port wing perhaps? *Newark Air Museum*

Although there was more than one farewell ceremony for the Lancaster in RCAF service, it seems that KB976 was *officially* the last Lancaster to fly with the RCAF during a last day ceremonial event in April 1964. Even after that date it appears one or two non-effective airframes remained stored until SOC in early 1965, either that or it was a case of the admin' catching up. KB976 is seen almost at the end of its working life sporting a 1960s colour scheme with its code reduced simply to '976'. It was still a member of No.408 Squadron, itself part of Air Transport Command whose title appears on the aircraft's nose. Retired in April 1964, KB976 entered the Canadian civil register in June 1964. It was bought by the Strathallan Collection in 1975 but later sold on, only to be badly damaged when a hangar collapsed on it at Woodford on 12 August 1987. Surviving parts from this aircraft currently reside in Florida, USA.
Tony O'Toole collection

Who(se) nose ? Prior to its departure to the Strathallan Collection in Scotland, G-BCOH (ex-KB976), had its modified and extended nose replaced by a 'standard' unit from a long-redundant privately-owned Lancaster. The nose transparencies seen here belong to a Mk.10AR, so presumably they were taken from KB976's extended nose and grafted onto the donated unit.
Newark Air Museum

A reasonably detailed photograph of KB839, the third member of the Mk.10AR trio, taken while in preservation. As with KB882 and KB976, this Lancaster was also operated by No.408(P) Squadron until it was taken out of service in 1961 and stored at Ontario until being selected for preservation in 1964.
Newark Air Museum

develop the Lancaster 10MR, many of which were redesignated Mk.10MP from 1955/56 onwards upon receipt of an APS-33 radar in a thimble-like dome in lieu of the older H2S system.

Sources vary on the precise number of Lancasters actually *employed* by the RCAF during the post-war years, but not by much. Most seem to agree on an overall figure of perhaps 100 to 105 Lancasters having been retained following their respective modifications which ranged from minimal to extensive according to function. That these quantities are reasonably close is surprising given the perplexing contradictions and confusions that thrive solely (it seems) to frustrate would-be authors, a problem that has continued to grow over the decades driven by the many sub-Marks that materialized – several of which are not even accepted in some quarters. Further uncertainty arises because a number of airframes were converted or redesignated, officially or unofficially, more than once – an element that can always be relied upon to provide confusion.

With these considerations in mind it is hoped that some readers might find the tables on pages 42 and 43 useful.

Arguably the most colourful Lancasters ever seen were the two drone carriers KB848 and KB851 which, following conversion, became the Mk.10DC. Both had operated against Germany and were placed in storage following their return to Canada in June 1945. Subsequently selected for conversion, both were sent to Fairey Aviation, Nova Scotia, in January 1957 to become 'mothers' to Ryan Firebee radio-controlled target drones. Both Lancasters had launch racks, wiring and control systems installed for the 1,000 lb thrust, 600mph jet-powered drones that were intended to provide target facilities for Canada's later generation of interceptors.

Strictly speaking the Mk.10DC didn't serve the RCAF *per se*, as they were operated primarily by the Central Experimental & Proving Establishment (CEPE) whose unit code 'PX' is seen on the fuselage of 'PX-848' (KB848). After having flown to the US Naval Auxiliary Air Station, Brown Field, California, in order to confirm the feasibility of using Lancasters to launch this particular type of drone, the pair were mainly employed at RCAF Cold Lake, Alberta, until their involvement in the Firebee trials ceased in 1961. Thereafter KB851 was disposed of on 28 August 1961, while KB848 was SOC in early April 1964. (The CEPE was formed on 1 September 1951 by amalgamating the Experimental & Proving Establishment, the Winter Experimental Establishment [WEE], and other units, and by 1957 its HQ was at RCAF Uplands, Ontario, which possessed the longer runways required by CEPE's jet-powered aircraft.)

Clearly, both KB848 and KB851 made extensive use of high-visibility Fluorescent Red-Orange anti-collision paint, slightly less obvious was the fact that the drones and the Lancasters' spinners were red. From June 1958 (1953 for European-based RCAF aircraft) the Red Ensign began to replace the long-standing fin flash – but change could be a protracted process and it is likely that neither received the Red Ensign. The black and white images were taken in 1959/60; while the colour images date from 1960/61.

This page: *B&W images Newark Air Museum.* **Following page:** *colour image (port-side close up) via Tony O'Toole: other colour images are Author's collection.*

THE LANCASTER IN RCAF SERVICE

Lancaster 10MR, KB959 'AF-A', No.404(MR) Squadron RCAF. The Mk.10MR was a development of the Mk.10BR and although the latter was a fairly basic conversion it did receive rear-fuselage observation windows, rear-facing cameras, de-icing boots and a later H2S radar fit. Provision was also made for the carriage of bomb bay fuel tanks and depth charges. The more advanced Mk.10MR incorporated radar and sonobuoy crew stations in the fuselage for which the mid-upper turret was removed and the crew complement increased to ten. The code 'AF-A' on the upper wing is of interest as the practice of applying wing codes was meant to cease in early 1949. However, No.404 Squadron didn't reform until 30 April 1951 and was the first unit to which KB959 was allocated since being stored in 1945. Thus it seems likely that the practice of applying wing letters in full continued into 1951 in this instance. The Squadron used 'AF' until November 1951 when it switched to 'SP'. *Author's collection*

Above: KB959 'AF-A'. Below the nose turret are the characters 'A59' denoting the aircraft's identifying letter plus the last two digits of its serial number. Given that the wing letters appear on the top of the wing, it is likely that the full code 'VC-AFA' appeared below. Beneath the cockpit the squadron badge can be seen, while on the tailplane fin flashes appear on both the inner and outer fin surfaces, which was not always the case with RCAF Lancasters. Ultimately KB959 was later upgraded to Mk.10MP standard and remained in use until SOC on 17 May 1963. *Brandon White*

Below: Lancaster 10MR, KB974 'QT-974', No.121 Composite Unit (KU) RCAF. Converted in 1950, this aircraft's service history notes only that it served with No.404 (MR) Squadron in 1952 and that it was SOC in June 1955. Obviously KB974 also served with No.121 Composite Unit (unit code 'QT'). Although undated the code system had by this time altered whereby the aircraft's identifying letter on the fuselage was replaced by the three digits of its serial number. *Author's collection*

Lancasters 10MR, KB973 'AJ-973', and, KB892 'AJ-892', from No.407(M) Squadron. KB973 had previously been one of the thirteen Mk.10BRs and would later be upgraded again to Mk.10MP standard. It was SOC in September 1960. KB892 remained with this unit for the remainder of its service life having progressively received the codes 'AJ-D', 'AJ-892' and later 'RX-892'; it was SOC in June 1960. Both aircraft have the last three characters of their respective serials repeated beneath the nose turret and both have underwing roundels. *Brandon White*

Lancaster 10MP, FM213 'CX-213', of No.107 Rescue Unit seen at RAF Prestwick, Scotland on 19 June 1957. Previously a Mk.10MR the presence of the thimble-shaped radome housing an APS-33 radar in lieu of the earlier H2S readily marks this aircraft as a Mk.10MP, a system that began to appear on maritime Lancasters from 1955 which ultimately helped to prompt the new classification. FM213 was retired in April 1964 and SOC on 30 June 1964. It was then stored, seemingly with both turrets still in situ. Restored to flight in Canada, on 8 August 2014, FM213 landed in the UK despite the appalling summer weather that affected much of the eastern half of England that day. *Newark Air Museum*

Lancaster 10MP, KB889 'XV-889', seen post-1955 with both turrets in situ whilst serving with 2 (Maritime) Operational Training Unit (2MOTU), this image being one of the few to illustrate a Mk.10MP as a dedicated maritime patrol Lancaster as opposed to a S&R aircraft. *Brandon White*

Subsequently withdrawn from maritime patrol duties, KB889 is seen whilst serving with No.107 Rescue Unit based at Torbay, Newfoundland, where it arrived in June 1960. A comparison between the two images reveal several distinct changes in KB889's appearance, as well as a code which has been reduced to simply '889'. No.107 RU appears to have been the last unit with which this aircraft served and in 1962 (or April 1963 – sources conflict) KB889 was placed in store until SOC in 1964. Today this Lancaster resides in the UK at Duxford.
Newark Air Museum

THE LANCASTER IN RCAF SERVICE 37

Lancaster 10MP, FM104 'CX-104' No.107 Rescue Unit. There is some speculation that FM104 was used for a time in the S&R role from late 1945, which, if correct, presumably means it was flown in an essentially unmodified condition. In any event FM104 was subsequently stored until it was selected for conversion to Mk.10MR standard in April 1951, and a few years later was upgraded to become a Mk.10MP. This aircraft had a long and distinguished career which included the monitoring of Soviet vessels during the Cuban missile crisis before finally being retired in April 1964 and subsequent preservation.

Because of FM104's longevity it arguably became the most photographed of all of the RCAF's S&R Lancasters. However, the markings which accompanied Lancasters in this role altered, often quite subtly over the years, which is the point of this selection of images. Assuming a picture is worth a thousand words, then several thousand are saved here by allowing readers to study the changes for themselves.

Below: This, the earliest photo in the sequence, was taken in 1958. As related elsewhere 'CX' refers to No.107 Rescue Unit as witnessed by the legend below the nose turret. *Author's collection*

Above: The second of the sequence, taken at Prestwick in 1959 – no nose turret now and Red Ensigns have replaced the earlier fin flash. *Roger Lindsay*

Left: The third image in the sequence dates from 1960/61, 'somewhere in the UK' judging from the HP Victor beyond. *Tony O'Toole collection*

Above: This image illustrates FM104's final external appearance as seen at RAF Cottesmore in Rutland on 15 September 1962 being trailed by an Avro Vulcan. Whereas the red underwing panels extended only as far inboard as the outside edge of the control surfaces, above the wing the red panels extended further inboard to stop just short of the RCAF roundel; the wing walkway lines were also red. Additionally, the upper surfaces of the tailplane matched those of the underside in being solid red (excluding the control surfaces). *Tony O'Toole collection*

Below: Also taken at Cottesmore on the same date, this image is included to provide a starboard view of FM104. *Author's collection*

Top: Lancaster 10P, FM217 'VC-AKR', No.408 Squadron whose badge is seen below the cockpit. The earliest photograph in the Mk.10P sequence, FM217 displays the early post-war era codes with 'AKR', outlined in white (where the code overlaps the high-visibility red) clearly displayed beneath the port wing; the remaining portion, 'VC', appeared beneath the other. 'AK' represented the unit, while 'R' was the aircraft's identifying letter. Ten Mk.10Ps were converted, all had their armament removed and turrets faired over with the nose fairing incorporating distinctive portholes. The H2S radar was also removed while several new fuselage windows were added as were de-icing boots, several cameras and additional fuel tanks (in the bomb bay). The Mk.10N navigation trainer was derived from the Mk.10P and both shared a generally similar external appearance. FM217 served with No.408 Squadron until it crashed on 2 June 1960. *Brandon White*

Centre: Lancaster 10P, FM207 'MN-207' of No.408 Squadron in 1958. This, the seventh Mk.10P conversion, had previously been coded 'AK-207' until 1951 when the squadron code changed to 'MN' which was retained into 1958. During 1958 the RCAF dropped the use of squadron codes and it is likely that FM207 became 'RCAF 207' or simply '207' until SOC on 26 September 1962 for spares recovery. *Author's collection*

Above: Lancaster 10P, FM122 'MN-122' of No.408 Squadron seen at Woodford Airfield, Greater Manchester with an Avro Vulcan B.2 in the distance. This photo is believed to date from 1959 when FM122 visited the UK. It was SOC in 1962. *Author's collection*

Above: Lancaster 10P, FM122 'MN-122' of No.408 Squadron seen at RAF Scampton in 1959. *Author's collection*

Below: Lancaster 10S KB944. This aircraft was reportedly the least modified RCAF Lancaster in post-war service having received only minor modifications in 1950 prior to joining No.404 Squadron. Returned to storage in January 1957, KB944 was selected for preservation and in June 1964 (when photographed) it arrived at Rockliffe, Ontario, with guns reinstated prior to being placed in the National Aviation Museum. At this time KB944's outer wing panels, horizontal tailplane (top and bottom – excluding control surfaces) and propeller spinners were red with original-style fin flashes retained. *Newark Air Museum*

Just when you think that you've gained an understanding of RCAF colours and codes and that nothing could possibly shake your smug sense of wellbeing about the subject…this turns up! The point of interest here is the code, written in full *on the fuselage*. It hardly matters that the image quality is too poor to distinguish the serial number of this Mk.10N (or 10P) from 1 Air Navigation School (code 'DH') or that the latter didn't apparently operate Lancasters! *Author's collection*

Lancaster 10P, FM199 'MN-199' of No.408 Squadron offering clear close-up images of the Mk.10P. With regard to the photo taken from directly beneath the old bomb aimer's position, the point of its inclusion is that it can be seen (through the transparency) that FM199 had four portholes in the nose fairing where others appear to have had just three! Converted from July 1950, this Lancaster had previously been coded 'VC-AKF' and 'AK-199': it was SOC on 2 June 1960.
All Newark Air Museum

■ CONFIRMED RCAF LANCASTER 10 SUB-VARIANTS & ROLES

Mark	Role/definition	Alternative definition(s) employed by somesources (if applicable)
Lancaster 10BR	Bomber/Recce/Search & Rescue (S&R)	Bombing/Reconnaissance
Lancaster 10MR	Maritime Reconnaissance, S&R	
Lancaster 10MP	Maritime Patrol, S&R	Maritime Photographic
Lancaster 10DC	Drone Carrying	Drone, Drone Controller, Drone Carrier
Lancaster 10 O	Orenda jet engine development	
Lancaster 10P	Photographic Reconnaissance	Photographic Survey. Patrol (some sources appearing to confuse 10P with 10MP [Maritime Patrol!])
Lancaster 10N	Navigation Trainer	
Lancaster 10AR	Reconnaissance	Area, Aerial, Arctic Reconnaissance
Lancaster 10S	Standard	Stock, Storage, in Store (virtually unmodified)

Table 1 Notes

The Mark suffixes shown in the left-hand column are agreed upon. The middle column lists the correct definitions as determined by a majority of sources, while the third offers alternative interpretations from others. The authors leave readers to draw their own conclusions, however, it is worth mentioning that with the possible exception of the Mk.10MP, and pedants apart, none of column three's definitions materially conflict with those of column two.

The sub-types excluded from the table but which nevertheless are still to be found in some literature, old and new, covering post-war Canadian Lancasters are: **Lancaster 10SR; Lancaster 10S&R; Lancaster 10SAR; Lancaster 10ASR; Lancaster 10U; Lancaster 10T; and Lancaster 10C.**

The first four Marks all refer to the search-and-rescue role, but none were officially approved. The use of a search-and-rescue (S&R) suffix probably began with the Lancaster 10BR which, as indicated, was one of its defined roles anyway – as it was with the later Mk.10MR and 10MP too. Given their extended range, increased crew complements and later radar fits, a number of maritime Lancasters were allocated directly to S&R units despite which they retained their original suffix e.g. Mk.10MP. (It should be noted that 'non-maritime' Lancaster 10P and 10N variants were also allocated to the S&R role when required.) Additionally, from 1952 (probably earlier), Lancasters operating with S&R units began to appear with the word 'Rescue' painted in red on their bomb bay doors. Later 'Rescue' would appear on the fuselage sides enhanced by a fluorescent red/orange band edged in blue which remained until the last S&R Lancasters were retired in early 1964, by which point the RCAF must have squeezed virtually every ounce of life left in their last remaining examples.

In retrospect, given these considerations, it would be more surprising if unofficial S&R suffixes hadn't emerged, particularly from the later 1950s as the availability of the Mk.10MP for the S&R role inevitably increased as they were gradually replaced as maritime patrol aircraft on the West Coast by more modern and suitable types. (The last maritime patrol Lancaster was withdrawn from that role in 1959.) However, obsolescence to one side, the Lancaster remained relevant as a S&R (and reconnaissance) platform until finally the task of acquiring spare parts became a source of ever-increasing concern which led ultimately to the aircraft's withdrawal.

The **Lancaster 10U** (Unmodified) may never have existed yet the term survives tenaciously in some quarters where it is said to represent those aircraft which were retained or stored unmodified until disposed of – mostly between 1946 and 1950 as stated earlier. Conceivably the term Mk.10U might have been coined by bureaucrats or service personnel as an administrative measure merely to avoid confusion with other Lancasters that were to be retained: most particularly with the 10S (see table). Some sources seem to have entwined the Mk.10U with the Mk.10S making them one and the same, which was not the case. That the 10S existed, albeit in small numbers, is not in doubt though the precise quantity – from four to nine airframes – is. Any modifications made to the Mk.10S were minimal unless of course one was adapted for another purpose, of which Lancaster 10S KB917 is an example, having been used in the development of the Mk.10P although in the event KB917 was SOC on 2 January 1947 due to its poor condition. It appears that the Mk.10S existed to perform various miscellaneous flying duties as required, a purpose which, though limited, is sufficient to distinguish a Mk.10S from a Mk.10U (assuming the latter existed). Reportedly, of the Lancasters employed by the post-war RCAF, Lancaster 10S, KB944, was the least modified of all. Having received some minimal alteration in 1950, KB944 was used by a number of units until it was returned to store in January 1957 where it remained until May 1964 when it was flown into preservation in much the same condition as it had been when built almost 20 years earlier.

Only one very minor reference to a **Lancaster 10T** has been found by the author who fell upon it while researching other sub-types. In the absence of any further information whatsoever it is presumed that the reference was a typographical error.

The designation **Lancaster 10C** has been used in reference to FM205 which was used as a test bed to carry two Avro Canada Chinook turbojets in the outer engine nacelles, though it never actually flew in that condition. FM205 lingered until scrapped in 1956 so conceivably, given the designation Mk.10 O was applied to the solitary Orenda test bed FM209*, it seems likely that the Chinook test bed would have been referred to as the Mk.10C.

---

* See *Avro Lancaster, Lincoln and York In Post-War RAF Service 1945-1950*. Dalrymple & Verdun Publishing, for further Lancasters including test bed FM209.

■ RCAF LANCASTER SERIAL NUMBERS BY SUB-TYPE (WHERE KNOWN)

Mark	Notes
Lancaster 10BR	13 produced: KB907, KB919*, KB925*, KB946*, KB957*, KB961‡, KB965‡, KB973*, KB995†, KB996*, FM221, FM222‡, FM228†.
	Airframes marked † later modified to Mk.10MR. Airframes marked * modified to Mk.10MR and later to Mk.10MP. Airframes marked ‡ possibly modified to Mk.10MR – sources conflict!
Lancaster 10MR/10MP	Some sources suggest up to 70 MR/MP conversions: a quantity that would need to include all 13 Mk.10BRs, whereas it's probable that only 8 became MRs. By excluding the 10BR (and the further confusion created when 10Ps and 10Ns were used for S&R purposes), it seems likely that about 54 further Lancasters were converted to 10MR/10MP standards for a total of 62.
Lancaster 10DC	KB848, KB851.
Lancaster 10 O	FM209.
Lancaster 10P	10 produced: FM120, FM122, FM199, FM207, FM212, FM214, FM215, FM216, FM217, FM218. Initially seven examples were converted, subsequently three more Mk.10Ps were ordered. (FM208 might also have been a Mk.10P until becoming the Mk.10N pattern aircraft).
Lancaster 10N	5 produced: KB826, KB986, FM206, FM208, FM211.
Lancaster 10AR	3 produced: KB839, KB882, KB976.
Lancaster 10S	7 (but probably more): KB739, KB781, KB801, KB884, KB917, KB944, FM148.
	(KB739 used by Central Experimental & Proving Establishment (CEPE) for winter trials in 1945/46; SOC 1948.
	KB781 served until 1955 – last unit was No.408 Sqn; SOC 1/1956
	KB801 operated as a training a/c in fighter affiliation and naval exercises; SOC 8/5/56.
	KB884 & KB917 were both used to assist in the development of the 10P from late 1945, both returned to store in 1946 and 47 respectively due to their worn-out condition and disposed of.
	KB944, possibly the least modified of any post-war Canadian Lancaster; 11/5/64 preserved.
	FM148 used by CEPE for winter trials, later stored; SOC 9/1/58.)

■ RCAF LANCASTER SQUADRON CODES 1951 - 1958

Code	Squadron or Unit	Role	Date from	Date to
AF	404 Sqn	Maritime patrol	April 51	November 51
AG	405 Sqn	Maritime patrol	March 50	November 55
AJ	407 Sqn	Maritime patrol	July 52	1955
AK	408 Sqn	Photo-reconnaissance	January 49	1951
AP	413 Sqn	Photo-reconnaissance	May 47	November 51
CH	103 S&R Flight	S&R	1950s	?
CJ	123 Rescue Flight	S&R	Late 1940	?
CQ	Central Navigation School		August 1951	Mid-1950s
CX	107 Rescue Unit	S&R	1952	1958
DJ	2 Maritime Op Training Unit [2 (M)OTU]		Early 1950s	
FC	Central Exp & Proving Establishment [CEPE]		September 1951	?
GS	Central Flying School		1951	?
MN	408 Sqn	Photo-reconnaissance	Late 1951	1958
PX	Central Exp & Proving Establishment		1958 or 1959	
QT	121 KU (Composite Unit)	S&R	1954	1959
RX	407 Sqn	Maritime patrol	1955	1958
SP	404 Sqn	Maritime patrol	November 1951	1955
XV	2 Maritime Op Training Unit		November (?) 1951	1958

Table 3 Notes

The post-war RCAF initially coded their aircraft using a five-letter scheme known as the VC system. The first two letters 'VC-' represented the RCAF, the next two referred to a particular squadron or unit with the fifth being an aircraft's specific identifying letter. Lancaster lower wing surfaces were meant to display the full code 'VC-AFA' (for instance), the two blocks separated by the fuselage, reading left to right when viewed from below with the base of each letter bordering the ailerons. Underwing roundels were excluded. On the upper wing surfaces and fuselage sides only 'AF-A' would be applied. Needless to say variations and interpretations occurred with elements of this system lingering for years – and in some instances was still being applied – after the VC-system had been replaced.

In the early 1950s a new, simplified, two-letter system was introduced whereby each squadron or unit displayed only their specific two-letter code followed by the three numerals of an aircraft's serial number. The now unique code (two letters – three numerals) was to be applied on each side of the Lancasters' fuselage. This system functioned into 1958 after which the two code letters began to be replaced with 'RCAF' for a time until eventually Lancasters could be seen using just their three serial numerals as a code.

The table above lists those codes and units which are known to have had Lancasters on strength during the 'two-letter' period.

The Lancaster in Aéronavale Service

Lancaster WU-01 (previously B.7 NX613), probably seen in late 1951 without unit codes and with exhaust shrouds that do not appear to have been fitted to other French Lancasters. WU-01 was damaged during landing on 25 April 1954 and SOC fourteen months later.
Newark Air Museum

Following agreements made by the Western Union (a precursor to NATO) in 1948, arrangements were made to supply the *Aéronavale* (French naval air arm) with fifty-four ex-RAF Lancasters modified for the maritime reconnaissance role, but due to the rundown of the labour force within Britain's aero industries work could not commence until 1950. The aircraft, a mixture of B.1(FE)s and B.7(FE)s, were modified by Avro at their Woodford and Langar facilities, although the term 'modified' was an understatement as many were practically rebuilt and all had their dorsal turret removed. Numbered WU-01 to WU-54, the first (WU-01) was delivered to France in January 1952, and while most would serve abroad a number were retained in mainland France.

In 1952, five further Lancasters were ordered for the *Le Service Générale a l'Aviation Civile et Commerciale* (SGAC) by the French Government. These, as with the other French Lancasters, were thoroughly overhauled and modified and received the identities FCL-01 to FCL-05 (French Civil Lancasters). They were intended primarily for the dedicated search and rescue role along the North African coastline and were delivered between early January and late April 1954.

Although thirty-two Lancaster B.1s and twenty-two B.7s were earmarked for transfer, with lists of serial numbers published (married to their respective WU serials), it should be borne in mind that the initial airframe selection was provisional. The point being that there is evidence to suggest that one or more of the Lancasters initially chosen were not in fact sent, possibly because, on closer inspection, some were deemed unsuitable for modification and replaced by other stored Lancasters without amending the original list. The same applies to the FCL group all of which were to have been B.7s, however, at least one was replaced by a B.1.

French Lancasters were operated by several *Aéronavale* units including: *Flotilles* 2F, 10F and 11F (which, following a reorganisation in June 1953, became 23F, 24F and 25F respectively); *Escadrilles* 4S, 5S, 9S, 10S, 23S, 52S, 55S, 56S, 58S and 62S. The SGAC Lancasters were frequently, if not exclusively, operated by 24F, 25F and 55S in the S&R role until these Lancasters were withdrawn in 1960.

For ease of reference the following photos are arranged by ascending WU number.

Lancaster WU-04, 55.S 5 'E' (previously B.7 RT682), date and location not known. It was still in service in Morocco as late as September 1961. In comparison with other *Aéronavale* examples WU-04's fuselage and underwing roundels are of smaller diameter. *Author's collection*

Lancaster WU-08, 55.S 8 'H' (previously B.7 NX703) in company with two unidentified Lancasters. Based at Agadir in Morocco, *Escadrille* 55S was the multi-engine conversion unit for a number of types, including WU Lancasters from 1952 to 1961/62 and SGAC Lancasters until 1960. The unit moved to Corsica in 1961, after which its remaining Lancasters were steadily withdrawn. Although this image is undated it is known that WU-08 was damaged beyond repair in September 1958. *Newark Air Museum*

Lancaster WU-08, by now 55.S 1 'A' seen at Le Bourget, France, on 15 July 1957. Because of the date it is presumed that this is the later of the two WU-08 images, but it is only a presumption. A close inspection of the original photo shows that some of the paint on the numeral '8' has peeled away. *via Simon Watson*

Above: Lancasters at Agadir, Morocco in March 1953. As luck would have it 55.S 6 'F' is devoid of a WU number, either that or some colour other than white was used to apply it. About thirty-four *Aéronavale* Lancasters passed through the hands of *Escadrille* 55S over the years. *Newark Air Museum*

Below: Lancaster WU-14, 24F '5' (previously B.7 NX623). In 1952 Lancasters arrived at Lanne-Bihoué, Brittany, the home of *Flotille* 10F, although within a year the unit was renumbered to become *Flotille* 24F and continued to operate Lancasters into 1958. The latter's badge can be seen on WU-14's nose which, though hard to identify, comprises a stylized bird flying over either a lighthouse or a buoy. WU-14 was SOC in April 1961. *Newark Air Museum*

THE LANCASTER IN THE AÉRONAVALE SERVICE 47

Lancaster WU-16, 9S '1' (previously B.7 NX622). *Escadrille* 9S was formed at Lanne-Bihoué in 1957 with three specially prepared and overhauled Lancasters for use from French bases in New Caledonia in the South Pacific; all were painted white. The three; WU-16, WU-27 and WU-41 were replaced in 1962 by three further white-painted 'tropical' Lancasters; WU-13, WU-15 and WU-21, and in the event *Escadrille de Servitude* 9S (Surveillance) became the last French unit to operate Lancasters. The image seen here is a pre-1961 photo, the year that the old unit codes were replaced by identification numbers – thus 9S '1', seen here, became simply '16'. These Lancasters were periodically dispatched to Australia for major servicing and overhaul and when WU-16 was sent there for the last time in 1962, there it stayed and is today preserved near Perth.

Lancaster WU-17 (previously B.1 TW655) at Langar in early 1952 in far better condition compared to the photo seen in 'Second-line Lancs Miscellany'. This aircraft was sent to Port Lyautey, Morocco, home of *Flotille* 2F, the first French unit to use operational Lancasters which in turn replaced Wellingtons. WU-17 didn't last long – it was SOC in October 1953 following a heavy landing.

Lancaster WU-27, 9S '2' (previously B.1 TW651) is seen in Fiji in 1959/60 with earlier style unit markings which includes the number '2' at the top of the fin as well as on the nose.

Lancaster WU-27, 9S '2', a starboard view. *All Newark Air Museum*

Above: An unidentified Lancaster seen at a location unknown to the authors. Other photos exist that were taken at the same location and the same spot, a place where Lancasters were often broken up. However this ex-24F Lancaster (the unit badge is clearly visible) is not being scrapped but undergoing a major overhaul to the extent that the blue paint is being stripped off. Perhaps this is one of the six Lancasters that were allocated to the Pacific. *via Simon Watson*

Right: Of the six Lancasters sent to the Pacific, four survived into preservation: WU-13 (NX665, New Zealand), WU-15 (NX611, UK), WU-16 (NX622, Australia) and WU-21, although for the latter (previously B.7 NX664) salvation commenced some twenty years after the type was finally withdrawn. On 21 January 1963, this Lancaster crashed at Mata Utu in the Wallis Islands while landing on a grass airstrip – and there it remained. By May 1984, when these photos were taken, ultimately successful efforts were underway to salvage the remains and today WU-21 resides at Le Bourget, Paris, the only *Aéronavale* Lancaster to be preserved in France after earlier efforts to preserve WU-22 (on Le Bourget's fire dump) failed in the 1960s. *Newark Air Museum*

Opposite page:
WU-27, 9S '2' seen, this time, at Eagle Farm Airport, Brisbane, Australia in 1959. WU-27 was SOC in June 1962 and cannibalized for spares. *Author's collection*

Lancaster WU-40, 24F '6' (previously B.1 PA432). *Flotille* 24F continued to use Lancasters into 1958 when they were replaced by Lockheed P2V-7 Neptunes. WU-40 flew on however, to serve with a host of other units prior to being SOC in June 1961. *Newark Air Museum*

Maintenance in Morocco on Lancaster 'H' belonging to *Escadrille* 55S, which is all we know about this image! *Newark Air Museum*

Above: Almost authentic (if one ignores sponsors' nose art and civil registrations) is ex-*Aéronavale* WU-15 (previously B.7 NX611). Seen at Biggin Hill in May 1965, this aircraft had been extensively overhauled in Australia prior to its return flight to the UK which commenced on 25 April and was completed eighteen days later. To WU-15 goes the distinction of being the last Lancaster in French service (they were replaced by Douglas DC-4s), its last flight with the *Aéronavale* being made in August 1964 when it was delivered to Australia. *Author's collection*

Below: A 'nose-art' close up. The badge below the pilot's window is the *Escadrille* 9S motif. *Author's collection*

The Lancaster in other Air Forces

Argentinian Air Force service

Immediately after the end of the Second World War, the *Fuerza Aérea Argentina*, (FAéA, Argentine Air Force) began a process of modernisation, which included buying several British types such as the Gloster Meteor jet fighter, (and in doing so became the first air force in Latin America equipped with a jet-propelled aircraft) and a number of Avro Lincoln and Lancaster bombers, in an attempt to create a powerful strategic bomber force in the region.

In June 1947, the Argentine Air Force placed an order for fifteen stored RAF Lancaster B.1s that had been completed in the summer of 1945. They were taken out of storage and overhauled by Avro at Woodford and Langar. Originally in the PA and RA serial number ranges, the aircraft were given Argentine registrations numbering from B-031 to B-045 inclusive and were delivered between May 1948 and January 1949. The Lancasters were allocated to *Grupo 1 de Bombardeo*.

Lancasters remained in Argentinian service well into the 1960s, albeit in declining numbers and by early 1963 a nominal eight were left, the remainder having crashed, been otherwise destroyed, or grounded due to a lack of spares and engines. Even so it was the *Fuerza Aérea Argentina* which became the last military force anywhere to operate a Lancaster, the last flight of which occurred on 5 December 1965 when their last airworthy example, B-040, crashed at Rio Gallegos in Patagonia. Although three other Lancasters still remained none had flown for several months due to a lack of engines: they were officially written off in July 1966 and scrapped soon thereafter.

Royal Egyptian Air Force service

In the late 1940s, the British Government agreed to the sale of nine surplus Lancaster B.1s to the Royal Egyptian Air Force (REAF). Refurbishment of the airframes was completed at Bracebridge Heath, with all the Lancasters being taken from storage at Langar.

Following the completion of test flights in the UK, deliveries were ready to commence by November 1949, and continued into the second half of 1950. All nine Lancasters were officially received by No.8 Squadron, REAF, based at Almaza. Thereafter they appear to have been infrequently used, possibly due to a shortage of spares, and spent most of their operational life on the ground.

By the 1956 Suez Crisis, No.8 Squadron was in the process of re-equipping (as was No.9 Squadron, then equipped with the Handley-Page Halifax) and it is thought that

Lancaster G-11-14, temporary 'B-class' marks that allowed test flights to be conducted in the UK. Previously B.1 PA375, this Lancaster was to become B-031 in Argentinian service arriving in that country in September 1948. It was written off in 1962 and used as a source of spares. B-006 in the foreground is ex-RAF Lincoln RE353, one of thirty delivered to Argentina at more or less the same time as the Lancasters. RE353 was sold for scrap in 1967.
Newark Air Museum

Above: Lancaster B-033 at Langar in June 1948 leaving little ambiguity with regard to its identity. This was one of the three Lancasters still remaining with the *Fuerza Aérea Argentina* after B-040 crashed in December 1965, however, B-033 hadn't flown since the end of June that year owing to a lack of spare parts. *Newark Air Museum*

any remaining Lancasters were no longer operational, although they remained in what appeared to be a flyable state dispersed at various locations. As such they were considered too great a threat to be left alone and on 1 November 1956, Royal Navy Hawker Sea Hawks of Nos.804 and 810 NAS, from HMS *Bulwark*, attacked Cairo West airfield on a pre-dawn strike. Both squadrons' operational records report that one Lancaster was destroyed, another probably destroyed and one damaged. Other non-confirmed reports state that five or six Lancasters were destroyed on the ground at this time by the Sea Hawks although no other reference has been found to support such claims.

Sweden

Sweden obtained an ex-RAF Lancaster in 1950 for use as an engine testbed and in May 1951, B.1, RA805 arrived in Sweden suitably modified with a large underslung pod intended to accommodate a jet engine. In Swedish service the Lancaster was designated Type Tp 80 and given the identification number 80001; its first test with the Dovern jet engine commenced in June 1951. The Tp 80 was destroyed in a crash in 1956.

Below and opposite: Swedish Lancaster 80001 seen with a Dovern jet engine underslung. The underside of the fuselage was covered by a stainless steel covering to protect it from the jet's heat. *Newark Air Museum*

Post-war Lancaster Colour Schemes

Royal Air Force 1945-1956

Wartime Temperate Land Scheme with Night under surfaces

By the end of the war, the camouflage and markings which were applied to the Lancaster on the production line were well established, comprising the Temperate Land Scheme of Dark Earth and Dark Green on the upper surfaces, and Night (black) on the under surfaces, to what was termed Pattern No.2, meaning that the Night under surfaces extended high up the fuselage sides.

The Dark Earth and Dark Green disruptive camouflage pattern applied to the upper surfaces was in the 'A' Scheme, a specific pattern taken from Air Diagram 1161, titled 'Camouflage scheme for four-engined monoplanes – night bombers, torpedo bombers, general reconnaissance land planes, troop carriers and bomber transports'. Previously there had been 'A' and 'B' Schemes in Air Diagram 1161, which were mirror images of each other, to be applied to alternate aircraft on the production line, but in early 1941, the decision was made to ask aircraft manufacturers to select either one or the other, and Avro appear to have selected the 'A' Scheme, which was subsequently applied to every Lancaster which utilised a disruptive pattern on its upper surfaces on the production line.

At that time, national markings consisted of 84 inch diameter red/blue roundels on the upper surfaces of the wings and 54 inch diameter red/white/blue/yellow roundels, in the post-May 1942 proportions, on both sides of the fuselage. The red/white/blue fin markings were 24 inches high by 36 inches wide and applied to both the inner and outer faces of both fins. The red and the blue of these markings were in the dark mid-war shades and the yellow was a rich, egg yolk, shade. No national markings were applied under the wings. It is worth noting that the fuselage roundels were staggered so that there was room to apply the two-letter squadron codes forward of the roundel on the port side of the fuselage, so that the code and individual aircraft letter combination could be read from left to right on both sides of the aircraft.

The serial number was applied in red, 8 inch tall, characters high on the rear fuselage and following delivery to a squadron, the squadron codes and individual aircraft letter were applied to the sides of the fuselage, also in red. By the end of the war, No.5 Group had modified this practice by adding a thin yellow outline to the letters to make them stand out more clearly in daylight.

The 'daylight scheme' and the Lancaster B.1 Special

The 'daylight scheme' was applied to Lancaster B.1 Specials which originally equipped No.617 Squadron. However, following the decision to deploy No.617 Squadron as part of Tiger Force, the squadron relinquished its Lancaster B.1 Specials in June 1945 and the aircraft were passed to No.15 Squadron who flew them for some time after the end of the war for various bombing trials.

The colour scheme applied to these aircraft was the standard Temperate Land Scheme of Dark Earth and Dark Green on the upper surfaces, in the 'A' Scheme, with Medium Sea Grey under surfaces to Pattern No.1, which meant that the camouflaged upper surfaces extended down the fuselage sides.

The Tiger Force Scheme

Tiger Force, also known as the Very Long Range Bomber Force, was the name given to the British and Commonwealth long-range heavy bomber force formed in early 1945 from squadrons serving within RAF Bomber Command in the UK, for proposed use against targets in Japan. The unit was scheduled to be redeployed to the Far East in the lead-up to the Allies' invasion of Japan, but in the event, the force was disbanded after the bombing of Hiroshima and Nagasaki and the Soviet invasion of Manchuria which ended the war.

At some point towards the end of the European War, in late 1944/early 1945, when further thought was being given to the ongoing war in the Far East, a question was asked by Tiger Force Headquarters to HQ Bomber Command about the internal temperature of aircraft under tropical conditions. The temperatures inside an aircraft depended on the local weather conditions at any given time, but, it was estimated that an unpainted polished natural metal finished aircraft would, in full sunlight, usually be between 10 to 15 degrees centigrade cooler than a dark camouflaged painted one. Although there were disadvantages to not painting the upper surfaces of aircraft, including that on sunny days and moonlit nights the natural metal glint might give night fighters a greater 'pick-up' range, although on dark nights it was considered to be an advantage to have a light coloured upper surface as the bomber would be less visible. Natural metal finished aircraft were also more visible to aerial observation when parked at dispersal, and there was a risk of corrosion of the metallic surfaces, although it was felt that none of the disadvantages, except perhaps for corrosion, was very serious and for most purposes it was an advantage to have a light-coloured upper surface.

So, following practical experiments, it was agreed that whilst an aircraft with an unpainted aluminium surface had a lower internal temperature than that with a dark painted surface, the internal temperature of an aircraft painted white would be even lower, so white upper surfaces were adopted. It was also felt that searchlights were likely to be a serious menace in the Far East, and following research in the United States which indicated that a high-gloss black finish was more effective in delaying 'searchlight pick up' than a matt black finish, an operational trial had been conducted on a batch of fifty Rootesbuilt Halifax B.IIIs during late 1944. These trials were considered to be very successful and this led to the adoption of a gloss black finish for the under surfaces of RAF Heavy Bombers which was known as Anti-Searchlight Glossy Black and it was intended that this finish be applied to the under surfaces of Tiger Force aircraft.

Accordingly, by the end of June 1945, a scheme of matt white upper surfaces to Pattern No.1, and Anti-Searchlight Glossy Black under surfaces was promulgated for Tiger Force Lancasters, although it would appear that Anti-Searchlight Glossy Black under surfaces were not applied to the earliest aircraft produced for Tiger Force. Armstrong Whitworth began to make deliveries of the aircraft which would become Lancaster B.I(FE)s in June with conversion and final painting of these aircraft being carried out by Short & Harland in Belfast. It is thought that Armstrong Whitworth delivered the aircraft already finished in the standard wartime Temperate Land Scheme and Night finish and that as a result, although the upper surfaces were refinished in white, the standard Night finish was retained on the under surfaces.

The Austin Motor Company, who were also contracted to build Lancasters during and after the war, were also unlikely to have applied Anti Searchlight Glossy Black to their early production Lancaster B.VIIs, as the first Lancaster B.VII built by them, NX611, was delivered on 16 April 1945, before the 'black and white scheme' had been decided upon, with production being completed by September 1945. Therefore whilst Anti Searchlight Glossy Black might have been subsequently applied to the Tiger Force Lancasters, it is thought that few had it applied on the production line.

As for squadron markings, it was initially intended that aircraft in the Tiger Force Scheme should be marked with red squadron codes and individual aircraft letters – a strange decision in view of the arguably justified paranoia which surrounded the use of red in any kind of marking in the Far East that might be mistaken for the Japanese national *hinomaru* marking, but this was the standard colour used for this purpose by Bomber Command at this time. Red codes were applied to some Lancasters in the Tiger Force Scheme for a time, although it would appear that they ultimately gave way to the black codes introduced from November 1945. It is also thought that it was from about this time that the serial number on the rear fuselage also began to be applied in black, replacing the earlier red characters. Underwing serials were marked in white, approximately 48 inches high, reading from the front under the port wing and from the rear under the starboard wing.

The post-war grey and black bomber scheme

As the end of the war in Europe was in sight, proposals regarding a change in the camouflage of night bombers began to be considered as part of a review of existing aircraft camouflage policy which was being undertaken by the Air Ministry. In January 1945, the camouflaging of RAF aircraft was under review in the light of changing operational conditions and the easing in manpower and material which would result if aircraft were supplied in a non-camouflaged state. It was suggested that Allied air superiority was such that operating certain types of aircraft, uncamouflaged, was acceptable, although non-camouflaged aircraft, particularly the larger types, would show up very clearly on the ground.

It was stated that there would undoubtedly be a saving in material and manpower if aircraft were accepted without camouflage, and it was noted that a large number of USAAF aircraft had been in service, uncamouflaged, for some time, and the Air Ministry requested Bomber Command's views as to whether they were acceptable in this state from the operational and maintenance points of view.

Bomber Command was of the opinion that the application of the current Temperate Land Scheme of Dark Green and Dark Earth to the upper surfaces was a waste of time and labour. Apart from the work expended in laying out the pattern, units had to maintain supplies of the two colours and it would possibly be a great advantage if the two colour disruptive pattern could be dispensed with in favour of a single colour.

Some experimental work had been carried out by the Bomber Development Unit in 1943 which indicated that night bombers would be less easily picked up by night fighters if their upper surfaces were painted in a lighter colour, but nothing had been done to implement any change to the production camouflage scheme. It was considered unwise to abandon camouflage on any operational bomber while Bomber Command was still engaged on night operations, but it was suggested that it would be possible to dispense with the two colour disruptive pattern on the upper surfaces in favour of a single colour, which should be a dark grey or green, preferably with a gloss finish. It was also suggested that the under surfaces for heavy bombers should remain black, but here again, a glossy finish was recommended. All RAF Commands agreed that heavy night bombers should retain some form of camouflage finish and as Bomber Command was prepared to accept a single colour in place of the existing two colours on the upper surfaces of their aircraft, it would still result in a saving of manpower. However, in the event, it was not until after the war in the Far East had also ended that the Air Ministry was in a position to begin its review of aircraft camouflage policy.

A meeting was held in October 1945 to discuss post-war aircraft camouflage policy, at which plans were set for a period of five years, with the provision that they would become subject for review after four years. It was agreed at this meeting that all night bombers were to be painted with Dark Sea Grey upper surfaces and Anti Searchlight Glossy Black under surfaces, the boundary between the upper and under surface colours to be Pattern No.2. Spinners were to be Dark Sea Grey. For some reason it does not appear that the Tiger Force Scheme was discussed – let alone considered for retention. It didn't take long though for this camouflage scheme to run into trouble as by June 1946 the suggestion was already being made that the Dark Sea Grey finish on the upper surface be replaced by Medium Sea Grey, as the Dark Sea Grey was a very poor heat reflector in hot climates. Evidently this was found to be a valid point because by March 1947, the decision to replace Dark Sea grey with Medium Sea Grey had been taken.

By October 1948, all night bombers and bombers with both a day and night bombing role, were finished with Medium Sea Grey upper surfaces and

Anti-Searchlight Glossy Black under surfaces to Pattern No.2, with the equally proportioned, post-war bright red/white/bright blue national markings. These regulations then remained in force until No.214 Squadron, the last Lancaster heavy bomber squadron, finally gave up the type in March 1950.

Although comparatively few Lancasters wore the post-war Medium Sea Grey and Anti Searchlight Glossy Black scheme with the revised equally proportioned national markings in the bright colours, of those that did, some carried a variation in the colour of the national markings which was also seen on some of Bomber Command's Lincolns. In this variation, whilst the roundels on the upper surface of the wings were marked in bright red, white and bright blue, the fuselage roundels and fin markings only retained the bright red. The white was replaced by a glossy, slightly 'off-white' shade, whilst the blue was a comparatively light blue, which can best be referred to colloquially as a 'light bright blue', the reason for its use presumably being to make the roundel stand out more against the dark background.

Air-sea rescue and maritime reconnaissance

It would appear that all the Lancaster ASR.3s were produced as conversions of aircraft which left the production line as bombers. This conversion work was carried out by Cunliffe Owen Ltd at Eastleigh near Southampton over a two year period from July 1945 onwards.

In July 1945, the instructions for the application of the camouflage and marking of Air Sea Rescue aircraft were that the upper surfaces were to be Extra Dark Sea Grey and Dark Slate Grey and the under surfaces were to be Sky, to Pattern No.1. There was an exception to this whereby if specially required for aircraft destined for overseas, then the under surfaces could be Azure Blue. The Temperate Sea Scheme colours applied to the upper surfaces were to the same 'A' Scheme that had been used for the wartime Temperate Land Scheme, but with the Extra Dark Sea Grey taking the place of Dark Green and the Dark Slate Grey taking the place of Dark Earth.

National markings were to consist of red/white/blue post-May 1942 wartime-style roundels on the upper surfaces of the mainplanes and on the under surface of the wings, with post-May 1942 red/white/blue/yellow wartime-style roundels on both sides of the fuselage and the post-May 1942 red/white/blue stripes on both the internal and external faces of both fins. The serial number was applied in black 8 inch characters on the rear fuselage and under the mainplanes in black characters approximately 48 inches high, reading from the front under the port wing and from the rear under the starboard wing. The squadron code and individual aircraft code letters were to be red, although No.210 Squadron was using a light colour, probably white, on its ASR.3s during late 1946.

The roundels applied to the under surface of the wings appear to have been applied in a variety of sizes, sometimes unusually large, being somewhere in the vicinity of 72 inches diameter. Subsequently, possibly following repainting by the Service as part of the normal maintenance of the aircraft, smaller roundels were applied. Whatever their size, these underwing roundels were located under the wingtips, outboard of the serial number.

Photographic reconnaissance Lancasters

Compared with the sagas surrounding the Lancasters camouflage and markings in the Bomber, Air Sea Rescue and Maritime reconnaissance roles, the story of the colour scheme carried by the Lancaster PR.1 is relatively straightforward.

The Lancaster PR.1 entered service during the summer of 1946 with a Flight of No.541 Squadron which ultimately expanded into No.82 Squadron which then operated the type until December 1953. The other squadron to operate the type was No.683 Squadron which formed at Fayid in the Suez Canal Zone in November 1950 and then operated the type until November 1953. These aircraft were converted from standard bomber airframes and some of the PR.1 conversions are known to have been modified from aircraft finished in the Tiger Force Scheme described earlier, with post-May 1942 wartime national markings in the usual positions, which they initially retained.

At the October 1945 Air Ministry conference however, it was decided that all Photographic Reconnaissance aircraft were to be painted in a silver finish with the smoothest possible surface in order to produce their best performance. Even though Coastal Command successfully petitioned the Air Ministry to reintroduce an overall PRU Blue finish for PR aircraft in 1947, the overall painted-Aluminium finish remained in use on the Lancaster PR.1 for the whole of its service career which was spent almost entirely in the Middle East and Africa.

Their roundels remained in the usual size and location (although underwing roundels do not seem to have been carried), changing from the post-May 1942 wartime markings to the post-war markings as previously described in 1947. Serial numbers appear to have been applied in the usual places and sizes in black characters. From the few available photographs, it does not appear that squadron codes were carried by either unit although individual aircraft letters were sometimes applied to the sides of the fuselage in a variety of locations and colours.

One of the more noticeable features of the Lancaster PR.1 was the extent of the modification made to the rear of the cockpit canopy. Initially PR.1s simply had their Perspex 'roof' panels painted over as a temporary expedient to deflect the sun's heat. Later a metal fairing replaced much of the canopy's original transparent panels.

RAF Coastal Command

At an Air Ministry conference held in October 1945 to determine post-war aircraft camouflage policy for the next five years, the Air Sea Rescue category of aircraft appears to have been dropped. In the interest of standardisation, it was decided that all Coastal Command and Fleet Air Arm aircraft were to be Extra Dark Sea Grey on the upper surfaces and Sky on the under surfaces with the exception of what the minutes of the meeting describe as 'heavy types' of Coastal Command aircraft which were to have white under surfaces.

This new policy covering Coastal Command aircraft stated that all medium and long range anti-shipping, anti-submarine and general reconnaissance aircraft (except those made of wood and fabric) were to be finished in Extra Dark Sea Grey upper surfaces with white under surfaces, to Pattern No.1 – the under surfaces of the main and tailplane were to be glossy white. The under surfaces between the boundaries defined by Pattern No.1 and Pattern No.2, (i.e. the fuselage sides), were to be matt white. In addition, the matt white was to be extended upwards on to the upper surfaces in such a manner that in both front and side elevations, the aircraft would appear to be almost entirely white. In order to help facilitate this, the engine nacelle forward of the boundary of the upper surface colour on the wings was also to be white.

Although national markings as such were not really covered in this policy, Coastal Command aircraft were to carry the post-May 1942 style proportioned red/white/blue roundels on the upper surfaces of the wings and both sides of the fuselage. No roundels were to be carried under the wings. Fin markings were also not specifically mentioned, but it would appear that this was because it was tacitly understood that the post-May 1942 style red/white/blue marking would be applied. Serial numbers, both on the rear fuselage and under the mainplanes, were to be applied using the then new standardised characters in Light Slate Grey. Despite these instructions, that the underwing serial numbers were to be applied in Light Slate Grey, it would appear that many aircraft had them applied in Night, which might have been a result of misinterpreting the instructions.

It is thought that this was the scheme which was applied to the early Lancaster GR.3, such as those which went to equip No.210 Squadron – the first squadron to receive the type in June 1946. Although the colour of the squadron codes was supposed to be red, photographic evidence shows that on some occasions, other colours were actually used. Lancaster GR.3s of No.203 Squadron used Night codes whilst No.224 Squadron appear to have used red codes on their GR.3s during 1947. Some squadrons also appear to have used Medium Sea Grey whilst others did actually manage to use the correct colour, for example No.203 Squadron in 1950. The reasons for these variations in colour are not known and may simply have been the result of a shortage of the necessary materials necessitating the use of whatever was available in the immediate post-war years.

As mentioned above, by June 1946 Bomber Command was already suggesting a change from Dark Sea Grey to Medium Sea Grey on its bombers due to the poor heat reflection of Dark Sea Grey. This may also have been a concern for maritime aircraft, as by July 1947, the upper surfaces of medium and long range anti-shipping, anti-submarine and GR aircraft were now also to be finished in Medium Sea Grey – the white under surfaces as previously described were to be retained. The other alteration in the colour scheme of Coastal Command aircraft which took place at this time was the adoption of the 1-2-3 proportioned national markings using bright red, white and bright blue colours as previously described. Despite the fact that Coastal Command aircraft in the 'white scheme' were not supposed to have underwing roundels, many Lancaster GR.3s did have them applied, and whilst the size could vary, they were always applied outboard of the serial numbers.

This scheme then remained in use, unaltered, until the policy review of 1950. The review concluded that the present Coastal Command scheme seemed to be the best that could be devised and that it should be retained. No changes were therefore recommended.

New coding system

Although no change in the camouflage of Coastal Command aircraft took place during the early 1950s, the existing system of applying code letters did. On 20 February 1951, the Air Ministry wrote to all Flying Commands at home and overseas informing them that it had been decided to abandon the existing system of squadron codes with immediate effect, and in its place, a system was adopted whereby a single unit identification letter would be used to identify the unit on the station to which an aircraft belonged. Although the use of this unit identification letter was optional, and need not be applied, individual aircraft letters were considered necessary.

When used, the unit identification letters were to be placed aft of the fuselage roundel. The letters A, B, L and T were to be used as necessary for all operational units, and the letters C to K (with the exception of E and I) were earmarked for operational training and non-operational units of Bomber, Coastal, Fighter and Transport Commands. The letters M to Y inclusive, less T and V, were to be used by non-operational units of Home Command, Flying Training Command, Maintenance Command, 90 Group and Technical Training Command. The letters C to Y inclusive, (less E, I, L and V) were intended to be used by non-operational units of overseas Commands. The identity of an aircraft within a unit was to be established by the use of a letter or number. The use of a letter or a number was optional, but numbers were primarily intended for units with more than twenty-six aircraft. If letters were used, they were to be single letters. If numbers were used they were to be two figure numbers starting at 10. This letter or number was to be placed forward of the fuselage roundel.

As applied to Coastal Command's two remaining operational Lancaster squadrons, which were both based at St. Eval, No.210 Squadron used the letter 'L' until it gave up its Lancasters in late 1952 whilst No.203 Squadron used the letter 'B' on its Lancasters until March 1953. Both squadrons continued to use the same code letter on the Neptune MR.1s which replaced the Lancasters. The situation is less clear where Lancaster GR.3s were serving within commands overseas. Whilst the last operational Lancaster squadron in the RAF, No.38 Squadron, retained its Lancaster GR.3s in Malta until February 1954, it is not known what marking, if any, (given that the markings were optional), replaced the squadron's 'RL' codes.

The overall Dark Sea Grey scheme

Although there were no Lancaster GR.3s remaining in the frontline squadrons by the time that the overall Dark Sea Grey scheme was introduced in 1955, the School of Maritime Reconnaissance did have a number of Lancaster GR.3/MR.3s on strength, which had been repainted in the overall Dark Sea Grey scheme by the time the Lancaster was finally retired from RAF service in October 1956.

Royal Canadian Air Force

Royal Canadian Air Force (RCAF) bomber squadrons had participated in the war effort from 1941 and were initially attached to RAF Bomber Command Groups. Canada, however, wanted its own identifiable presence in Allied air operations overseas, and it did not want its air force to be merely a source of manpower for the Royal Air Force. To this end, No.6 (RCAF) Group was formed on 25 October 1942, initially with eight squadrons, but at the peak of its strength, the Group consisted of fifteen squadrons, many of them equipped with Lancasters.

Following the war RCAF Lancasters Initially retained their wartime Bomber Command Dark Earth and Dark Green Temperate Land Scheme upper surfaces with Night under surfaces and standard late-war RAF national markings, but during the later-1940s, all RCAF aircraft, including the Lancasters, were re-finished in an overall natural metal scheme, literally created by stripping all the paint from the airframes. A matt black anti-glare panel was invariably painted in front of the cockpit and during 1947 the RAF style national marking was replaced by the RCAF red maple leaf within a blue/white roundel in six positions. For several years RCAF Lancasters retained the RAF late wartime style red/white/blue fin flash, which was normally applied to both sides of each fin.

Two letter squadron codes with an individual aircraft letter were carried on the fuselage sides separated by the RCAF roundel. Below the wings a two-letter prefix VC code was applied under starboard wing, with the squadron code and individual aircraft letter under the port wing, all reading from the rear. Above the wings, the two letter squadron code was applied above the port wing, while the individual aircraft letter was applied above the starboard wing – both inboard of the roundels, and again, all reading from the rear.

Red, high visibility 'search markings' were applied to both surfaces of the wing tips, extending well inboard but painted around the roundels and not overlapping the ailerons, and on both surfaces of the tailplanes but not on the elevators. Occasionally, white was painted along the top section of the fuselage dorsal spine.

In late 1951, the individual aircraft letters were replaced by a three digit number system, based upon the last three characters of an aircraft's serial number, although the squadron code letters were retained. Red high visibility 'search markings' were retained on the wing tips and tailplanes, but all the lettering was removed from both surfaces of the mainplanes.

The next markings change came with the universal introduction of the Red Ensign to replace the late-war RAF style fin flash in 1958, which had initially been applied to RCAF aircraft operating in Europe from 1953.

During the late 1950s, white fuselage spine upper surfaces were introduced as standard, and a red flash with a white centre was applied along the length of the fuselage as a demarcation between the white spine and remaining natural metal fuselage. ROYAL CANADIAN AIR FORCE titling in red with a black drop shadow was applied above the fuselage flash on the side of the white spine. Squadron code letters were deleted during this period, with just the three serial number digits being retained on the fuselage sides, generally to the rear of the fuselage roundels. No codes were carried above the wings but RCAF titling was applied under the starboard wing with the three numerals of the serial under the port wing, all reading from the rear.

During the late 1950s and in to the early 1960s, the red, high visibility 'search markings' were changed to fluorescent red-orange, and reduced in size to generally just cover the wing tips on a line with the end of the ailerons, although the tailplanes remained fully covered (still with the exception of the elevators). Anti-corrosion grey paint was applied to the under surface of most of the remaining maritime reconnaissance aircraft, and aluminium paint was often applied to the engine nacelles and the areas above and below the mainplanes that were affected by the exhaust stains.

Additional markings in the form of RESCUE titling in fluorescent red-orange was applied to those remaining Lancasters still undertaking search and rescue duties, which was later enhanced by a broad fluorescent red-orange band around the rear fuselage edged in blue, which was truncated by the RESCUE titling. Essentially this was the final scheme and markings applied to the Lancaster in RCAF service when it was ceremonially retired from the RCAF at Downsview, Toronto, in April 1964.

Aéronavale

As mentioned, fifty-four Lancasters were converted to maritime reconnaissance aircraft for the *Aéronavale* by Avro under a Western Union agreement signed in 1948 with the aircraft being delivered from 1952 – originally to supplement RAF patrols over Atlantic and Mediterranean shipping lanes.

Initially the Lancasters were painted in a factory-applied overall dark royal blue scheme with *Aéronavale cocardes* applied to all six positions, each narrowly outlined in yellow with a black anchor superimposed. Blue/white/red fin flashes were applied to the four fin surfaces, each with a black *Aéronavale* anchor superimposed. The *Escadrille* and/or *Flotille* codes were usually carried on the fuselage sides, positioned in between the *cocarde* and the trailing edge of the wing. The WU (Western Union) serial number appeared below the outer fin flashes while the individual aircraft letter sometimes appeared above them – all in white. Unit badges and the individual aircraft letter were generally applied beneath the front turret.

An addition to the overall blue scheme was the occasional application of a longitudinal yellow band along the length of the fuselage, applied during the mid-1950s, which was not particularly standardised in its positioning – some aircraft featuring it high up on the fuselage and some lower down.

From early 1957, because of the intention to use the type in the South Pacific three *Aéronavale* Lancasters were repainted in an overall white scheme, often with a black anti-glare panel between the windscreen and the front turret and sometimes with stylish black paintwork along the engine nacelle sides to help mask the exhaust stains. *Aéronavale cocardes*, still with narrow yellow outlines and a black superimposed anchor, were carried in all six positions with the blue/white/red and black anchor fin flashes sometimes applied to the outer faces of the fins, but not always. Subsequently three further Lancasters were similarly modified and they replaced the first three in 1962.

From mid-1961 the *Escadrille/Flotille* codes were changed for the *Marine Francaise* individual aircraft numbering system, which were applied in black with USAF-style 45°corners. When this marking system was introduced, the individual aircraft number was positioned on the fuselage sides, in between the *cocarde* and the trailing edge of the wing, with just the individual aircraft letter on the outer fins – all in black.

The five SGAC Lancasters, FCL-01 to FCL-05, were also finished in overall dark blue, presumably the same shade as with the WU- Lancasters, although their national markings consisted of ordinary blue/white/red *cocardes* and fin flashes without *Aéronavale* yellow outlines or anchor. A longitudinal yellow band was painted along the length of the fuselage, interrupted mid-way by the letters 'SAR', applied in yellow, and with yellow wing tips (above and below) extending inboard to a position mid-span along the ailerons. Both rudders were yellow. At least one aircraft (FCL-01) had its fuselage spine painted white to help reduce the heat inside the fuselage, and others may well have had the same.

Argentinian Air Force

Although the post-war grey and black bomber scheme was being introduced at the time, it would appear that all the Argentine Lancasters were finished in wartime Dark Earth and Dark Green upper surfaces with either matt Night, or possibly Anti-Searchlight Glossy Black, under surfaces, to Pattern No 2. Propeller spinners were also Night or glossy black. Pale blue/white/pale blue roundels were carried above the wings and on the fuselage sides, with horizontally striped pale blue/white/pale blue fin flashes on the outer faces of the twin fins. The aircraft's serials were applied in large white characters on the fuselage sides to the rear of the roundel, and under the wings, reading from the rear under the port wing and from the front under the starboard wing, and repeated in smaller characters on the nose and sometimes on the outer faces of the fins above the fin flashes.

Egypt

The Lancasters were re-finished in the then current post-war RAF standard bomber scheme of Medium Sea Grey upper surfaces with Anti-Searchlight Glossy Black under surfaces to Pattern No 2 and Medium Sea Grey propeller spinners.

REAF green and white roundels encompassing a white crescent and three stars in the centre spot were applied in all six positions with green/white/green fin flashes on the outer faces of both fins. These Lancasters were allocated the construction/serial numbers 1510 to 1518 and individual aircraft identification numbers 1801 to 1809 applied in white Arabic numerals on the rear fuselage and under the wings, reading from the rear under the port wing and from the front under the starboard wing.

(Note. The national markings described here were in use from 1937 to 1958, the year that Egypt merged with Syria to form the United Arab Republic and their previously separate air force identities were combined to become the United Arab Republic Air Force.)

■ LANCASTER B.I BASIC DIMENSIONS

Wing span	102ft 0in
Length (trestled)	69ft 6in
Chord (root)	16ft 0in
Tailplane span	33ft 0in

Glossary and Abbreviations

A&AEE	Aeroplane & Armament Experimental Establishment
AFS	Advanced Flying School
ASR	Air-sea rescue
ASV	Air-to-surface vessel (radar)
ANS	Air Navigation School
APDU	Air Photographic Development Unit
ASWDU	Air-Sea Warfare Development Unit
BBU	Bomb Ballistics Unit
BCBS	Bomber Command Bombing School
CAW	College of Air Warfare
CCGS	Coastal Command Gunnery School
CGS	Central Gunnery School
CN&CS	Central Navigation and Control School
CNS	Central Navigation school
CPE	Central Photographic Establishment
CSE	Central Signals Establishment
EAAS	Empire Air Armament School
EANS	Empire Air Navigation School
EFS	Empire Flying School
FRL	Flight Refuelling Ltd
GR	general reconnaissance
H2S	An air-to-ground radar system
JASS	Joint Anti-Submarine School
MoS	Ministry of Supply
MR	Maritime reconnaissance
MU	Maintenance Unit
OCU	Operational Conversion Unit
OTU	Operational Training Unit
PR	Photographic Reconnaissance
RAE	Royal Aircraft Establishment
RAFFC	RAF Flying College
RCAF	Royal Canadian Air Force
RWE	Radio Warfare Establishment
SMR	School of Maritime Reconnaissance
SOC	struck off charge

Lancaster B.1, PA306 'BH-K', of No.300 (Masovian-Polish) Squadron RAF, based at RAF Faldingworth, Lincolnshire, 1946 | Finished in the standard wartime RAF Bomber Command scheme of Dark Earth and Dark Green upper surfaces with Night (black) under surfaces and wartime national markings. The serial number would have originally been applied in red 8in high characters but at some point had been repainted in white, probably in the summer of 1945 when white underwing serials were applied. Similarly, the previously red code letters were also repainted in white presumably at the same time. No.300 Squadron, a Polish manned unit, was disbanded in October 1946. PA306 was SOC in May 1947.

Lancaster B.1 (Special), PD127 'LS-S', of No.15 Squadron RAF, based at RAF Mildenhall, Suffolk, 1946 | PD127 was finished in the 'daylight bomber scheme' with the Dark Earth and Dark Green extended down the fuselage sides, as applied to the B.1 (Specials) which had originally equipped No.617 Squadron. They were subsequently passed to No.15 Squadron in June 1945 who flew them after the end of the war on various bombing trials. Under surfaces were Medium Sea Grey. Wartime national markings were retained with squadron codes in white as were the underwing serials, but the fuselage serials were still in red. PD127, shown carrying a 22,000 lb Grand Slam bomb, was SOC in October 1947.

Lancaster B.7, NX773 'FGG-C', 'Capella' of No.1 Central Navigation and Control School, RAF Shawbury, Shropshire, 1951 | Introduced in late 1948, the post-war bomber scheme originally used Dark Sea Grey upper surfaces, but as this colour was a poor heat reflector in hot climates, the decision was soon made to replace it with Medium Sea Grey, as illustrated here. Under surfaces were Gloss Black. Post-war equally proportioned, bright red/white/bright blue national markings were carried with yellow trainer bands around the rear fuselage and wings, which did not overlap the flaps. The three-letter Flying Training Command codes and individual aircraft letter were in white, as were the underwing serials but again the fuselage serial presentation was in red. An 8-pointed star and the name 'Capella' were carried on the nose in yellow.

Lancaster B.7, NX687 'FCX-R', of the Empire Flying School, RAF Hullavington, Wiltshire, 1948 | To help with the internal temperature of aircraft operating in hot climates, white upper surfaces were adopted, initially for Lancasters earmarked for Tiger Force for use against targets in Japan. Under surfaces were a high gloss, anti-searchlight black finish. In the event, Tiger Force was not used but Lancasters painted in the scheme were operated by various RAF units throughout the late 1940s, NX687 of the UK-based EFS being such an example.

By 1948, post-war equally proportioned, bright red/white/bright blue national markings were replacing the wartime styles, with black fuselage codes, introduced from November 1945. In this instance, the three-letter EFS code and the individual aircraft letter were grouped together to the rear of the fuselage roundels. The white underwing serials read from the front under the port wing and from the rear under the starboard wing. Yellow trainer bands were carried around the rear fuselage (obscuring the black fuselage serial number) and both wings. NX687 became 6816M in December 1950.

Lancaster ASR.3, RF325 'P9-J', of the Air-Sea Warfare Development Unit, RAF Ballykelly, Londonderry, 1949 | In July 1945, instructions were issued that air-sea rescue and maritime reconnaissance aircraft were to be finished in Extra Dark Sea Grey and Dark Slate Grey upper surfaces with Sky under surfaces, as illustrated by RF325. Wartime national markings were initially retained, those above the wings being red/white/blue as were those on the under surfaces of the wings. The fuselage serial number was supposed to be black, like the under wing serials, but in RF325's case appear to have been painted red. Squadron and individual aircraft code letters were also to be in red, although other colours, were used, such as light grey as in RF325's case.

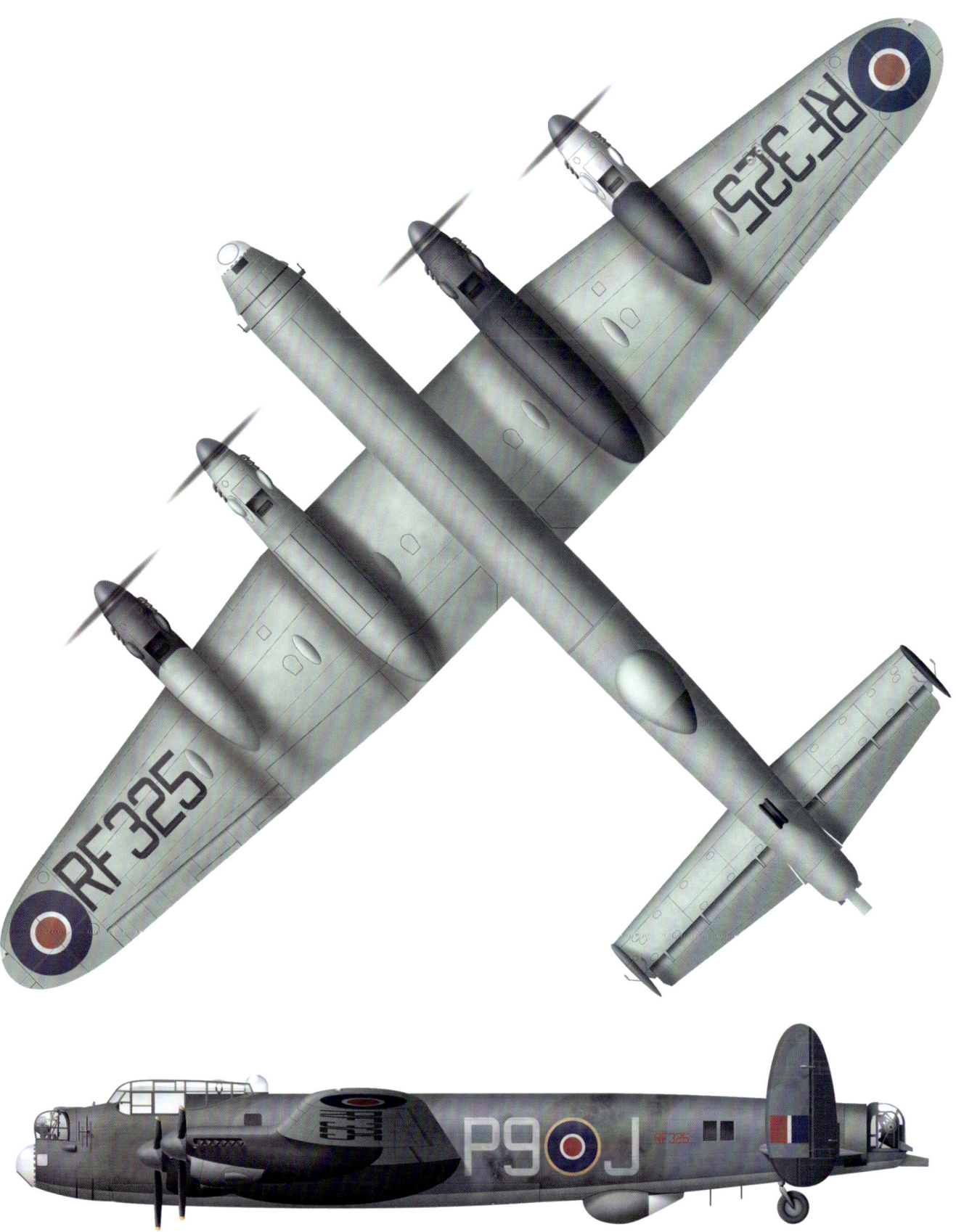

Lancaster ASR.3s were produced as conversions of standard production-line B.3 bombers by Cunliffe Owen Ltd over a two year period commencing in 1945. This particular airframe had seen arduous service both with the ASWDU and previously with No.279 Squadron, hence it exhibits very heavy exhaust staining and appears to have been fitted with various replacement parts from black as well as white coloured Lancasters.

Lancaster GR.3, RE186 'H-C', of the School of Maritime Reconnaissance, RAF St Mawgan, Cornwall, *circa* 1954 | By the end of the war, all medium and long range anti-shipping, anti-submarine and general maritime reconnaissance aircraft were finished in Extra Dark Sea Grey upper surfaces with white under surfaces. However, by mid-1947, presumably due to the same poor heat reflection of Dark Sea Grey experienced by Bomber Command, the upper surfaces of Coastal Command aircraft were also to be finished in Medium Sea Grey, although the tops of the engine cowlings on RE186 appear darker, possibly Extra Dark Sea Grey. The other alteration that took place at this time was the adoption of the 1-2-3 proportioned post-war national markings using bright red, white and bright blue colours.

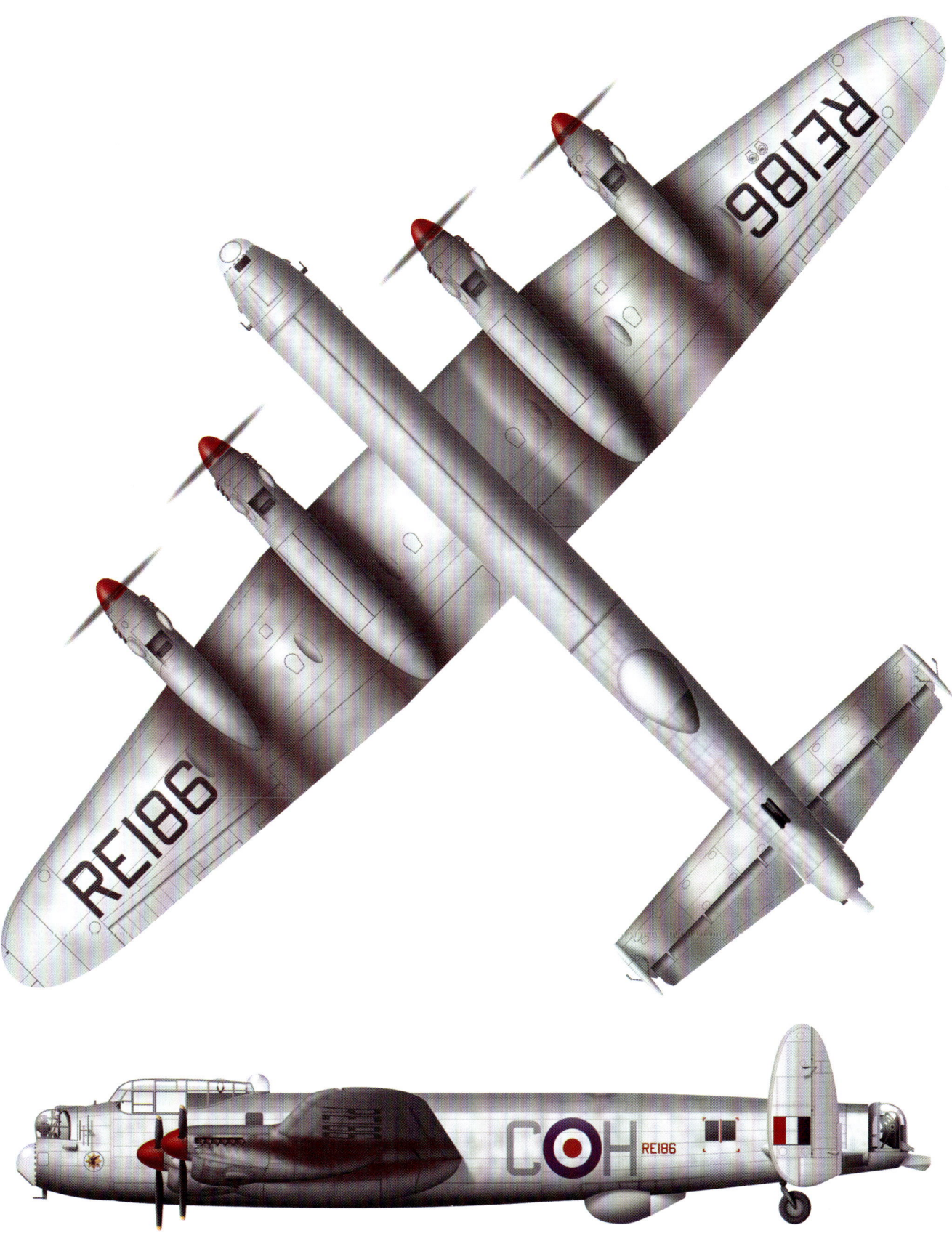

Serial numbers, both on the rear fuselage and under the mainplanes, were supposed to be Light Slate Grey but despite official instructions it would appear that many Coastal Command aircraft had them applied in other colours, such as red and black – as illustrated here on RE186. This may have been a result of misinterpreted instructions or simply a shortage of the necessary materials necessitating the use of whatever was available in the immediate post-war years. The unit's badge was applied to both sides of the nose and the propeller spinners appear to have been painted red.

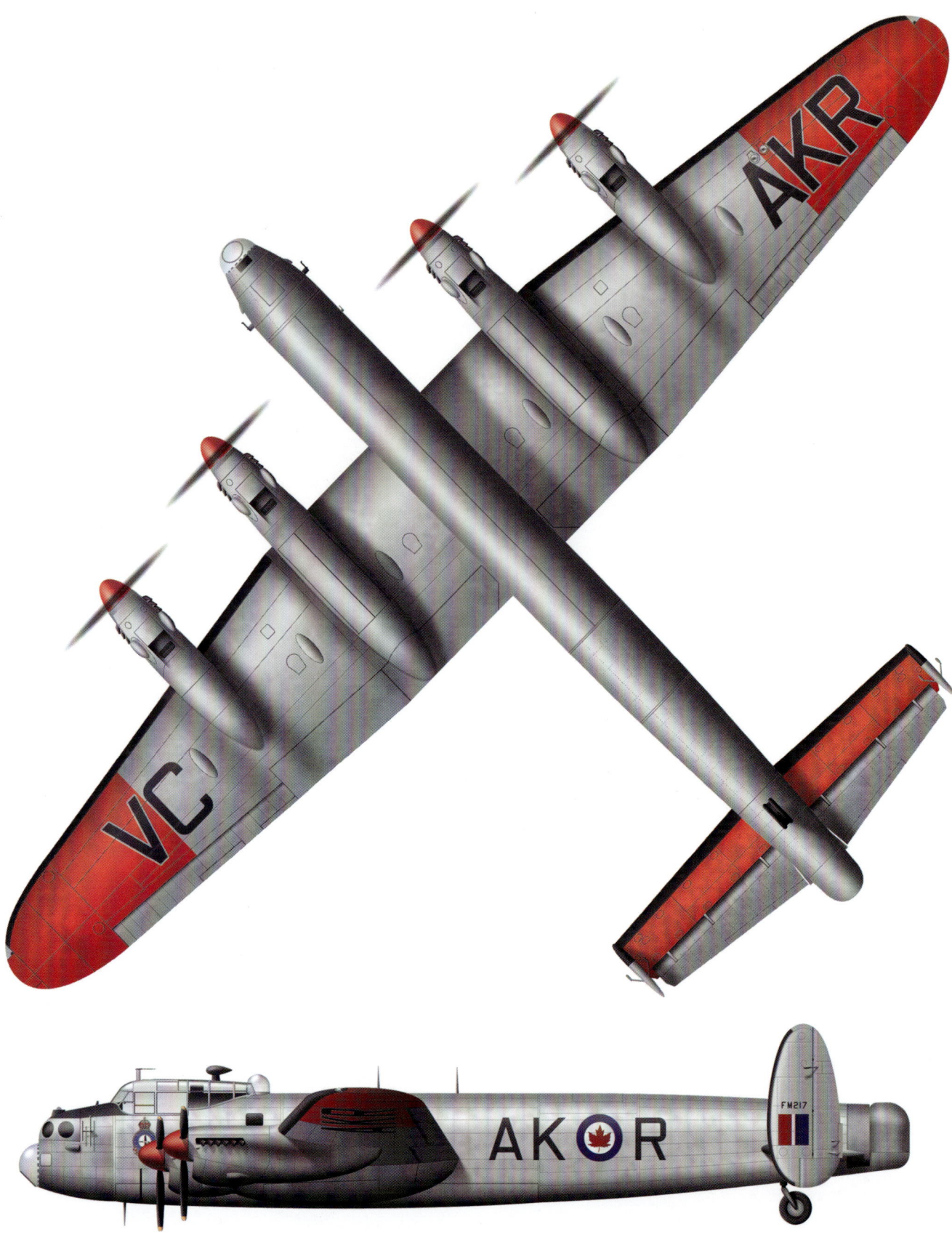

Lancaster 10P, FM217 'VC-AKR' of No.408 (Goose) Squadron RCAF, RCAF Station Rockcliffe, Ontario 1949 | During the late-1940s, all RCAF aircraft, including Lancasters, were re-finished in an overall natural metal scheme created by stripping all the paint from the airframes. RAF-style national markings were replaced by RCAF maple leaf roundels although the RAF style red/white/blue fin flash was retained on both sides of the fins. The squadron codes and individual aircraft letter were carried on the fuselage sides separated by the roundel, although below the wings the prefix 'VC' was applied beneath the starboard wing, with the squadron code and individual aircraft letter grouped together under the port wing. Red, high visibility markings were applied to the wing tips, extending well inboard but not overlapping the ailerons, and on both surfaces of the tailplanes but not on the elevators. Number 408 Squadron's Mk.10P and 10AR Lancasters were tasked with the mapping of Canada and it wasn't until February 1964 that they were finally retired and replaced with Dakotas.

Lancaster 10MR, KB959 'VC-AFA', of No.404 (Buffalo) Squadron RCAF, RCAF Station Greenwood, *circa* mid-1951 | No.404 Squadron reformed at RCAF Station Greenwood in April 1951 as a maritime reconnaissance unit equipped with Lancaster 10MRs. In similar fashion to FM127 (previously illustrated), KB959 was finished in the overall natural metal scheme with red, high visibility markings applied to the wing tips and tailplanes. On the wing upper surfaces the two-letter squadron code was applied to the port wing, while the individual aircraft letter appeared on the starboard wing – both inboard of the roundels. The characters A59 below the front turret represent KB959's individual letter and the last two numerals of its serial. Squadron badges were carried on the nose under the cockpit. In July 1951 this unit participated in Operation *Nanook'51*, to provide reconnaissance of the area between Resolute Bay and Thule, Greenland to look for ice.

Lancaster 10MP, FM213 'CX-213' of No.107 Rescue Unit RCAF, RCAF Station Torbay, Newfoundland, June 1957 | Even towards the end of its operational life with the RCAF, the Lancaster still provided valuable service – No.107 RU's role being to provide airborne search and rescue facilities and act as a homing beacon for overflying aircraft. FM213 was finished in the overall natural metal scheme with anti-corrosion grey paint on the fuselage under surfaces separated by a thin black cheatline. Roundels were carried in six positions, with the unit code 'CX' and the last three digits of the serial number either side of the fuselage roundel in black. During the late 1950s the red high visibility markings on the wing tips and tailplanes were changed to fluorescent red-orange, with a broad fluorescent red-orange band around the rear fuselage edged in blue. RESCUE titling was also added to the fuselage in fluorescent red-orange. FM213 looks very different today!

Lancaster B.1, B-038, of Escuadrón II, Grupo 1 de Bombardeo, Aérea Militar Coronel Pringles/Villa Reynolds, Fuerza Aérea Argentina, early 1950s | The *Fuerza Aérea Argentina* received fifteen ex-RAF Lancaster B.1s that were taken from storage and overhauled prior to delivery. The aircraft were finished in RAF wartime Dark Earth and Dark Green upper surfaces, black under surfaces, with *Fuerza Aérea Argentina* national markings and white serials. During its FAéA service, the Lancasters saw limited use while at least two are reported to have been converted into freight transports. Ultimately all were either written off or scrapped in Argentina.

Lancaster B.1, 1801, of No.8 Squadron, Royal Egyptian Air Force. Almaza, Egypt, mid-1950s | Nine Lancaster B.1s were taken out of storage, refurbished and delivered to the REAF between late 1949 and mid-1950. They were re-finished in the post-war RAF standard bomber scheme of Medium Sea Grey upper surfaces with Anti-Searchlight Glossy Black under surfaces. REAF green and white roundels were applied in all six positions with fin flashes on the outer faces of both fins. Individual aircraft identification numbers (1801 to 1809 inclusive) in white Arabic numerals were carried on the rear fuselage and under the wings. REAF Lancasters appear to have been infrequently used, possibly due to a shortage of spares, and spent most of their existence on the ground.

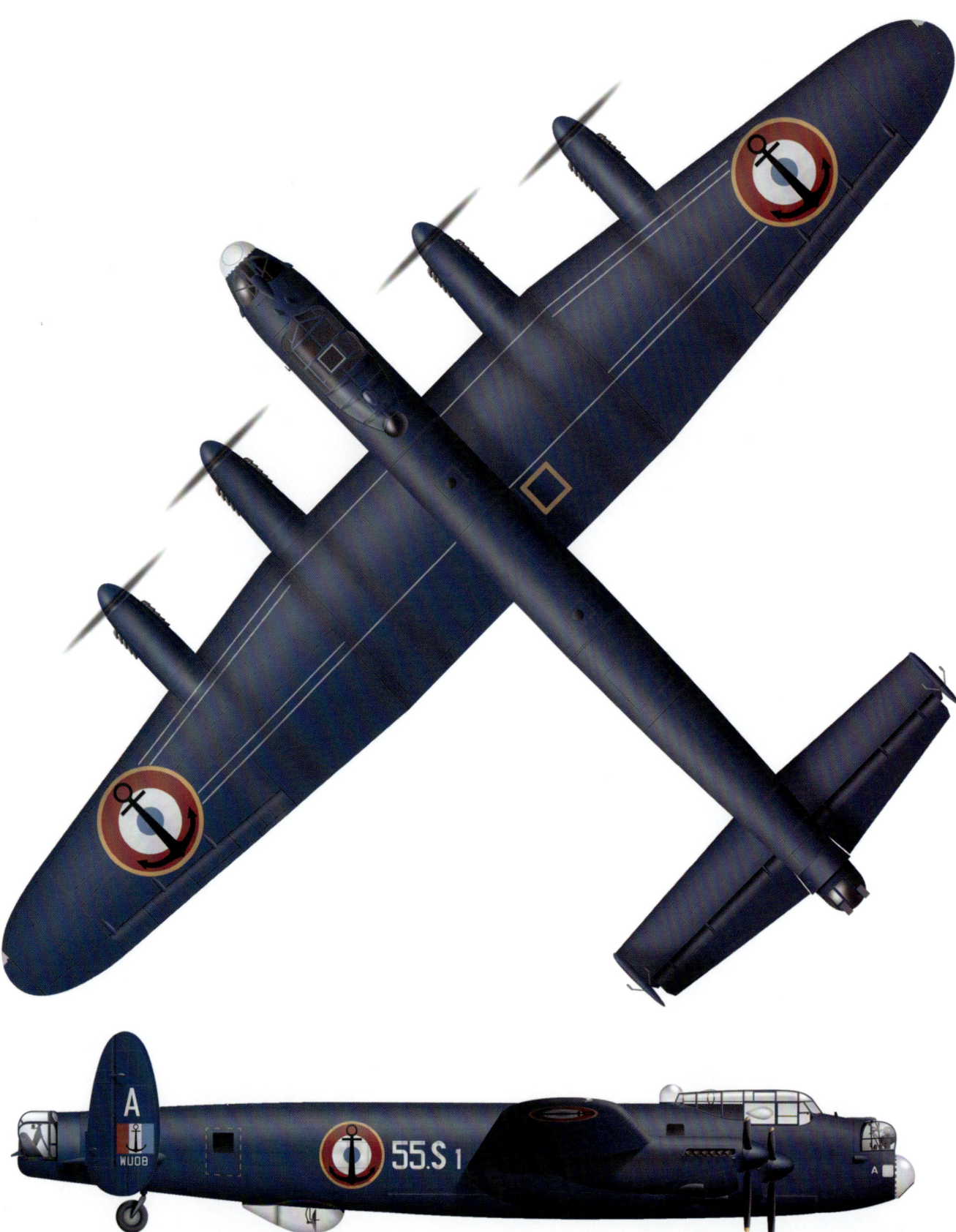

Lancaster B.7, WU-08 (ex-NX703), '55.S1/A' of *Escadrille 55S*, *Aéronavale*, BAN Agadir, Morocco, *circa* 1955 | Over fifty ex-RAF Lancasters were converted to Maritime Reconnaissance standard and delivered to the *Aéronavale* (French Naval Aviation) during 1952/53, of which over twenty were B.7s as illustrated here by WU-08. Initially the aircraft were painted in factory-applied overall dark royal blue, with *Aéronavale* markings and identification codes in white on the fuselage sides and outer faces of the fins. Based in Morocco, *Escadrille 55S* operated along the North African coast during the 1950s.

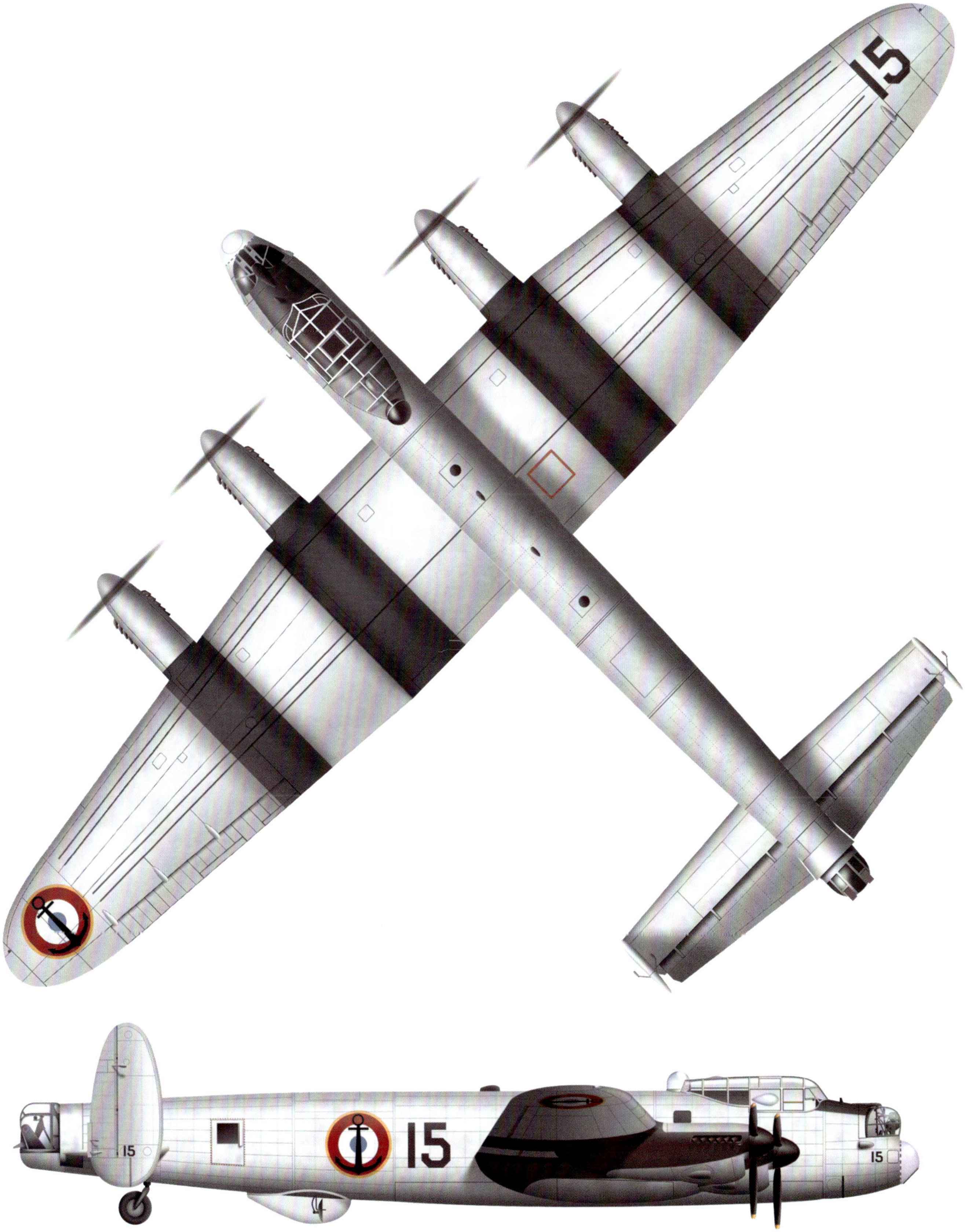

Lancaster B.7, WU-15 (ex-NX611), '15' of *Escadrille* 9S, *Aéronavale*, New Caledonia, southwest Pacific, 1962 | Three *Aéronavale* Lancasters were initially modified for service in the Pacific followed by three replacements a few years later. Because of the Pacific heat all six were repainted in an overall white scheme, sometimes with black paintwork extending along the engine nacelle sides and wing upper surfaces to help mask the exhaust stains, as illustrated here. *Aéronavale cocardes* were retained on the fuselage sides, but only carried above the port and below the starboard wings. The earlier style coding system was changed in mid-1961 and a single-number identification system introduced – applied above the starboard and below the port wings as well as on the fuselage. This particular aircraft, the last in *Aéronavale* service, is today preserved in the UK at the Lincolnshire Aviation Heritage Centre, East Kirkby.

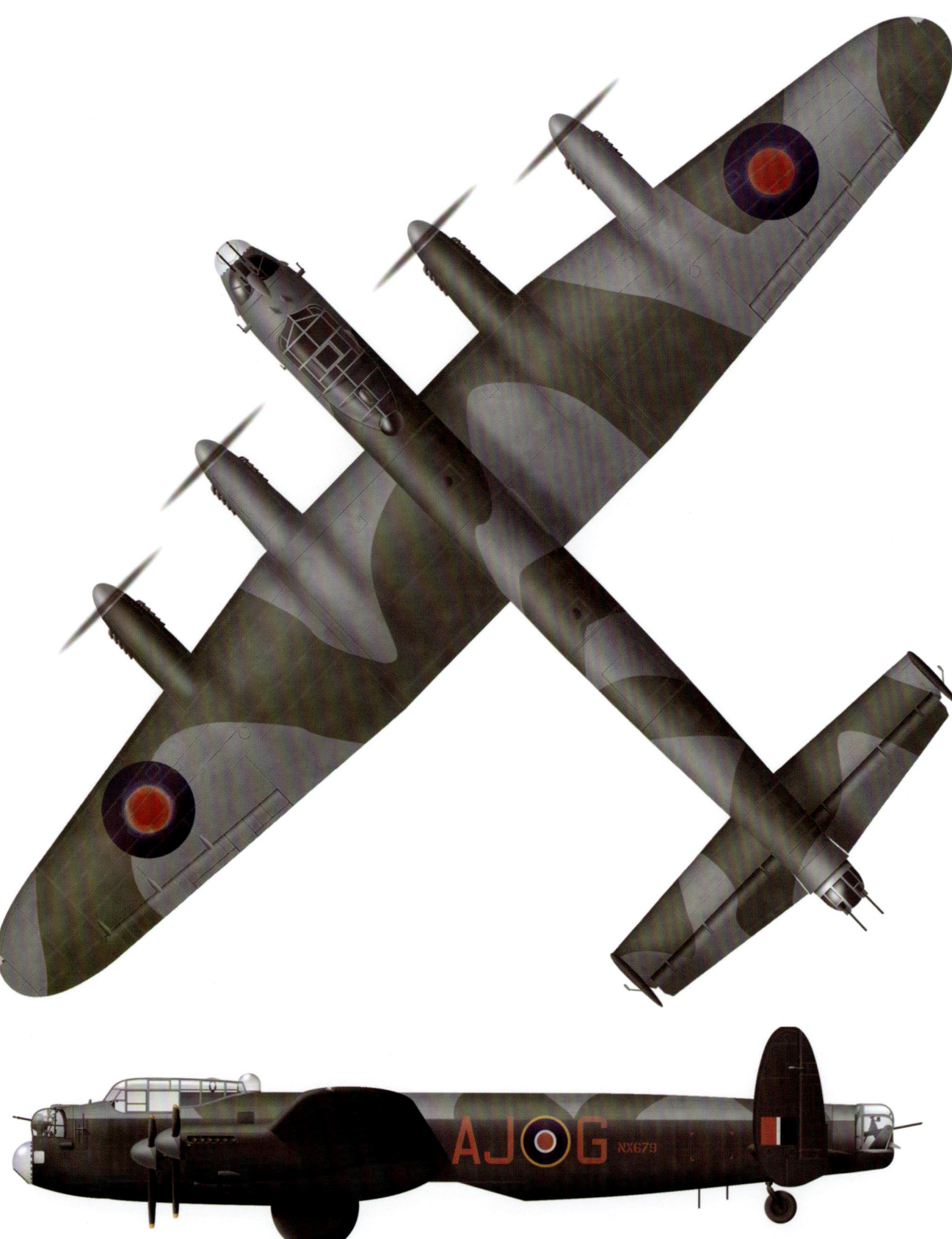

Lancaster B.7, NX679 'AJ-G', RAF Hemswell, 1954, used in the 'Dam Busters' film | Four Lancaster B.7s, fitted with FN.82 rear turrets, were used in the making of the 1954 'Dam Busters' film, NX679, illustrated here, being painted to represent W/C Guy Gibson's aircraft 'AJ-G'. It is possible that this was the Lancaster that had its serial temporarily altered to read ED932 (Gibson's aircraft) for a specific 'shoot'. Finished in the post-war bomber scheme of Medium Sea Grey and black, a dark green disruptive pattern was applied to the aircraft's upper surfaces to create a 'wartime bomber scheme', it not mattering that the colours were 'wrong' as the film was in black and white! The post-war national markings were also altered to match wartime ones and a huge dummy 'bouncing bomb' was fitted in the modified bomb bay. All four Lancasters were scrapped in July 1956.

Lancaster PR.1, PA474 'M' of No.82 Squadron, RAF Eastleigh, Kenya, circa early 1950s | No.82 Squadron reformed at RAF Benson in October 1946 equipped with Lancaster PR.1s and Spitfire PR.19s before moving to Kenya in October 1947 to undertake aerial surveys of Nigeria, the Gold Coast, Sierra Leone and The Gambia. Lancaster PR.1s were converted from standard bomber airframes and several were later painted in overall Aluminium with post-war national markings. One of the more notable features of the Lancaster PR.1 was the reduction of the cockpit canopy's rear glazing in an attempt to reduce the temperature inside. PA474 is now with the BBMF based at Coningsby in Lincolnshire.

Lancaster MR.3, RF325 'H-D' of the School of Maritime Reconnaissance, RAF St Mawgan, Cornwall, mid-1950s | Although no frontline Lancasters remained within Coastal Command squadrons by the time the overall Dark Sea Grey scheme was introduced in 1955, the SMR had a small number of Lancaster MR.3s on strength which had been refinished in the new scheme prior to the type being finally retired from service in October 1956, including RF325. Post-war national markings were carried with red fuselage serials. Codes and underwing serials were red outlined in white. The unit's badge was applied to the nose and it is thought that the propeller spinners were blue as illustrated.

Modelling the Lancaster

The Avro Lancaster has been kitted by most of the major injection-moulded plastic kit manufacturers over the last fifty years or so, invariably as Second World War Bomber Command variants, and many of these kits are still available. Most of the post-war operated variants were essentially similar to their wartime counterparts, and as such, many can be modelled with few changes from existing kits, although some featured improved or updated equipment and/or minor airframe changes that need to be considered.

MINICRAFT 1/144 scale

Avro Lancaster B.I

Originally produced by Crown in the 1970s, and then under the Academy label in the early 1990s, the moulds were acquired by MiniCraft in 2000. This was once the only option for anyone wanting to model a Lancaster in this scale

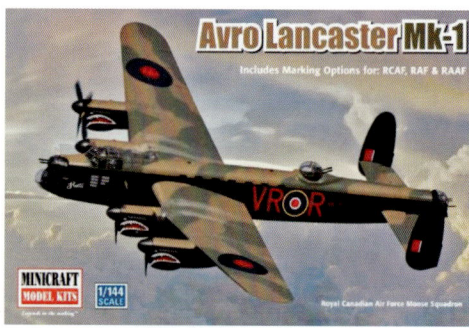

Three box tops under which the MiniCraft 1/144 scale toolings have been released over the years.

A-MODEL 1/144 scale

Avro Lancaster B.I/B.III

Never let it be said that 1/144 scale modellers don't relish a challenge – A Model products are 'limited run', so be prepared for plenty of flash and heavy sprue gates, but you also know that somewhere within the box there will always lurk a great model waiting to be liberated; and such is the case with the iconic Lancaster, a perfect subject for 1/144.

What you get are some 120 injection-moulded components which are certainly finer and better detailed than previous offerings from this manufacturer. An assortment of bombs for every occasion is provided (a chart included in the instructions suggests various possible combinations of load, albeit with no information as to their purpose). There are a number of alternative, duplicated or redundant parts such as exhausts, radomes and tailwheels, which presumably apply to the simultaneously released Canadian variant, (and possibly future versions?) A clear sprue covers the transparencies.

You also get a very delicate photo-etch set which provides for a complete bomb bay interior, radiator grilles, D/F loop, various mass balances and what appears to be the 'chain' used to hold a 'Tallboy' in place! The package is completed with some alternative mainwheels in vinyl, although the kit includes conventional plastic ones which might look better as 'under weight flats' can be filed on them, and a very passable decal sheet which offers two subjects – B.I, ME499, AS-D of No.166 Squadron RAF, and B.III, EE176 'QR-M' 'Mickey the Moocher' of No.61 Squadron RAF. The decals are very good quality but the blues and reds appear a little bright. Very fine upperwing walkway, and underwing trestle, lines are provided in black and red. A neat touch is the row of oval window decals along the fuselage. The shapes are lightly etched on the plastic and you can use them (or not) depending on the individual aircraft you are modelling.

The instructions indicate that you should start, logically enough, with the cockpit inte-

MODELLING THE LANCASTER 77

The A-Model Lancaster B.I/B.III finished in the kit's decals as a wartime B.I, ME499 'AS-D' of No.166 Squadron RAF, circa late 1944/early 1945 sporting yellow fins and wing tips. The small scale of the model is emphasized by the £1 coin, despite this there is plenty of detail, including quite a full bomb bay.
Model by Mike Verier

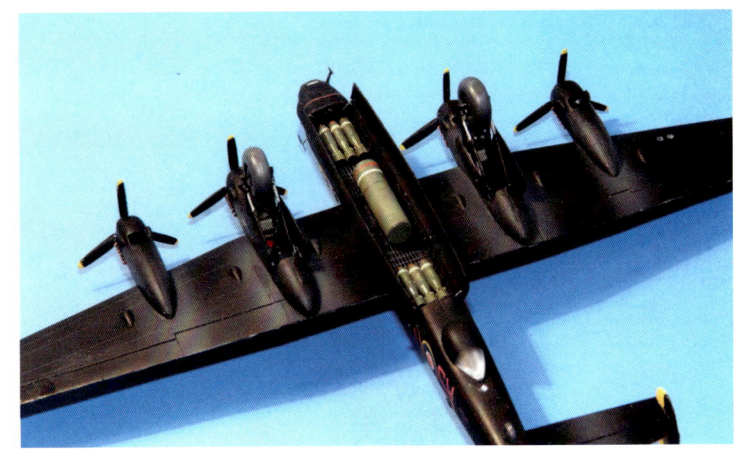

rior. This is all plastic and whilst the shapes are all there a fair bit of refining and fettling is required. You are also directed to build the etched bomb bay at this point – but don't. The photo-etch is exceptionally delicate and simply will not survive the handling required during construction if you try to put it all in at this stage. Fortunately the roof of the bay is a strong plastic component, essential for the structural integrity of the model.

The best approach initially is to just use the two large etched roof grids, everything else can go in later. Various transparencies also need to be put in before closing the fuselage halves, most of which don't fit very well, so again some patience and fettling is required. Once you have the thing assembled there is surgery to do. The main one being a large hole needed to accommodate the mid-upper turret. At this stage it also becomes apparent that the two circular transparencies on the centreline are too small. Fortunately, sections cut from the transparent sprue were just right so these were fitted and polished back to clarity. All the turrets can be installed later which greatly eases painting and handling.

There is no transparency for the mainplane leading edge landing lights, but these can be reproduced by simply burnishing down some kitchen foil on the inside, (a spot of adhesive will stop it slipping), before closing the wing around the two sturdy spars provided. Because of the spar design you can't really build the other wing until the spars are fed through the fuselage, although it might be possible to build the spars into the fuselage first. In the event, everything lined up okay and the finished result is quite robust. Before leaving the wings, wingtip navigation lights can be cut out and replaced with clear sprue which can then be polished back to clarity.

The engine nacelles are accurate but quite difficult assemblies having no locating pins. There are internal blanks and etched radiator grilles to contend with too. Again they go together eventually and fit the wings quite well. The kit provides alternative exposed exhausts (which are actually quite good) or rather vague lumps intended to represent the flame dampers.

The final phase covers the bomb bay, undercarriage, propellers (all with individual blades!) and the turrets. None of these are easy or quick to assemble, in particular the seemingly endless angle brackets for the bomb bay side walls. As an aside, the reason that the rear of the H2S radome was left unpainted was that it covered two of three downward i/d lights. There are a number of small aerials and details to fit before the bomb-doors can go on to complete the model.

Not for the faint-hearted perhaps, there is no doubt that this kit can be a challenging project, but ultimately a very rewarding one. The bomb-bay is stunning and proves that etched brass in this scale is certainly feasible. If anything, this model actually emphasises the need for a mainstream injection-moulded Lancaster and some decent decals in this scale, there are dozens of fascinating variants and colour schemes out there!

The kit has also been released as a Dambuster aircraft with the necessary mods and bombs. Bren Gun have produced a fine set of vacform transparencies as well as resin and photo-etch replacements for the bombbay and an excellent set of dropped flaps. Decals too have arrived from Kits World and Rocketeer, so it is possible to produce an excellent Lancaster in 1/144 scale now.

■ Assessment and model by Mike Verier

FROG

1/96 scale

Avro Lancaster B.I

The very first FROG (an acronym for Flies Right Off the Ground) kits approximated 1/96 scale, or thereabouts, and their Lancaster was one of the first kits released in the late 1950s. Festooned with heavily engraved panel lines and the positions of the markings suitably embossed, representing R5689 'VN-N' of No.50 Squadron, the kit was moulded in a glossy black plastic. Novo, Tri-ang and UPC released the kit for various period in the 1960s and 1970s and although

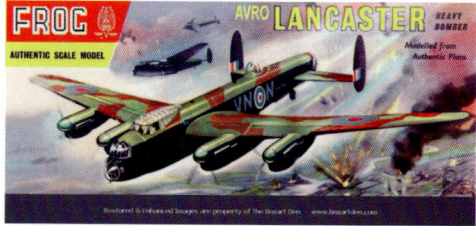

the kit is no longer currently in production or available, examples are sought after by collectors.

The original Frog 1/96 scale Lancaster kit box top from the late 1950s. Decal sheet markings represented R5689 'VN-N' of No.50 Squadron. The kit was moulded in a glossy black plastic. Novo, Tri-ang and UPC released the kit for various period in the 1960s and 1970s, and although the kit is not currently in production examples are sought after by collectors.

The colour scheme painting diagram printed on the Novo box underside.

AIRFIX
1/72 scale

Avro Lancaster B.I (first moulding)
Airfix's first 1/72 scale attempt at a Lancaster kit dates from 1958. Moulded in black plastic, and covered in raised rivet detail, it was generally accurate in its dimensions but had a couple of accuracy and outline issues. No longer available anyway, and replaced by much better examples, it is generally considered as a 'collectable' now.

Avro Lancaster B.I (second moulding)
In 1979 Airfix released a new tooling, which, when it first appeared was hailed as the best Lancaster kit available, although it was smothered in raised rivet and panel line detail, albeit somewhat more restrained when compared with their first moulding. Its outline shape and dimensional accuracy were good, although it was a relatively basic kit from the standpoint of the cockpit interior, gun turrets, mainwheel wells and bomb bay detail. The bomb aimer's blister was the short, early type and the propeller blades were the narrow versions, essentially

Left: One of the original box top presentations of the first Airfix 1/72 scale kit (top), with a subsequent boxing style (middle), and the final box top presentation before the second moulding kit was released (bottom).

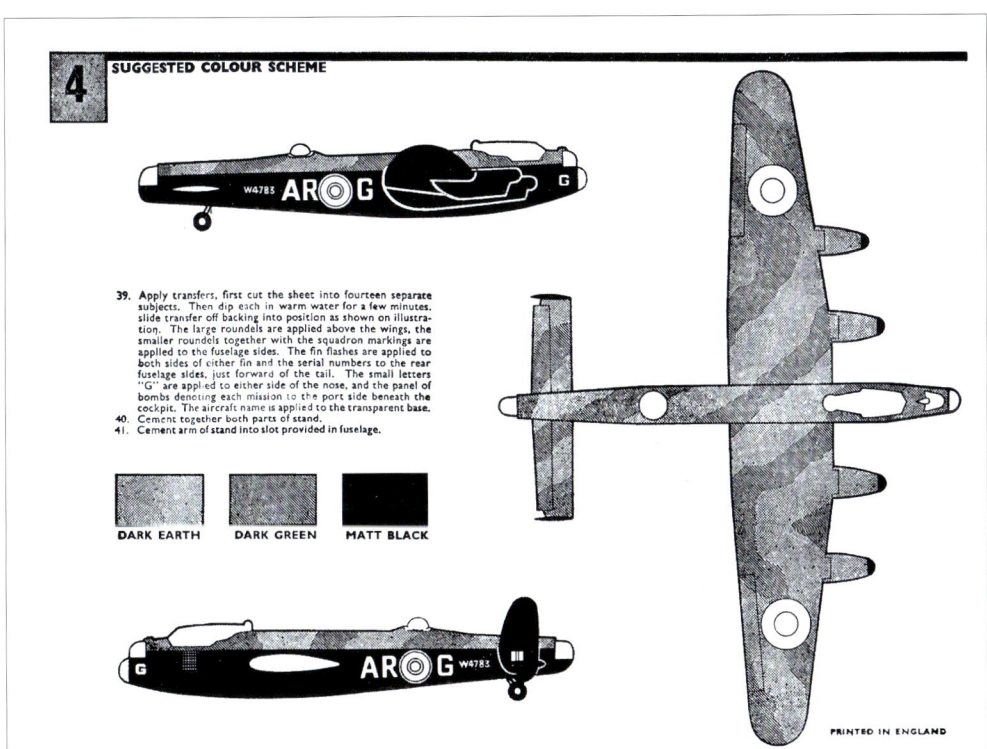

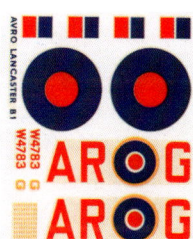

Left and above: The original 1958 painting instructions and decal sheet.

Below: The box top artwork and decal sheet circa 1979 of the second moulding release of a 1/72 scale Lancaster by Airfix.

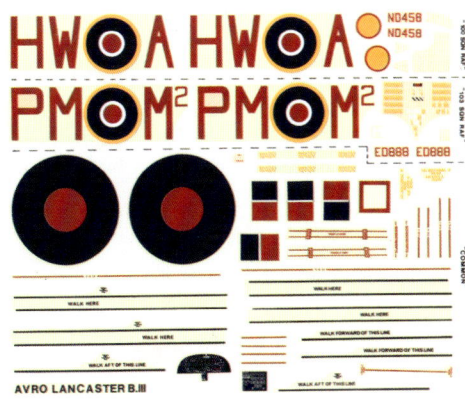

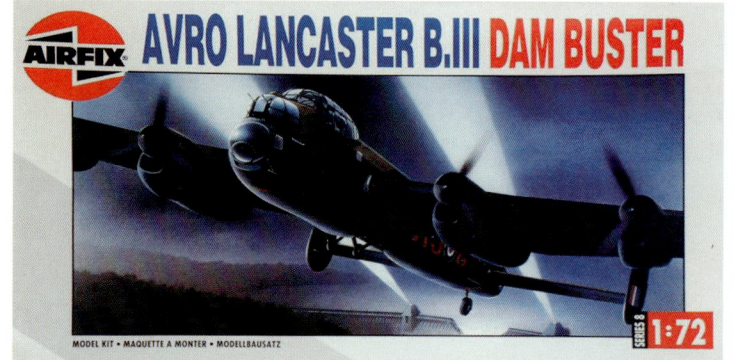
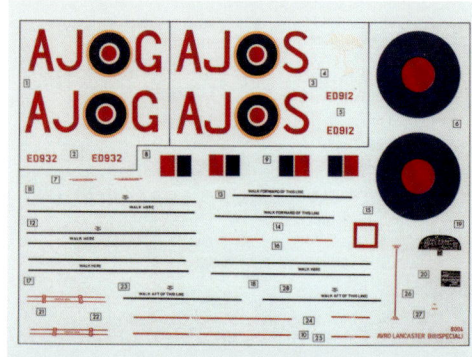

Top: The box top artwork and decal sheet for the second moulding release Airfix Lancaster re-tooled as a Dambuster for the 50th anniversary of the raid.

Above and right: The Airfix second moulding built as Lancaster GR.3, RE164 'H-U', in the markings of the School of Maritime Reconnaissance, based at St Mawgan in the early 1950s, in the white under surfaces with Medium Sea Grey upper surfaces scheme. Made from the Dambuster issue, search windows were cut out of the rear fuselage.
Model by Tony O'Toole

restricting it to representing an early B.I, although an H2S radome, in clear plastic, was included, and the instructions made reference to modelling it as a B.III too. There was a 'full' bomb load included, comprising a 4,000 lb 'Cookie' as well as 500 lb bombs.

Fit of some of the parts left a little to be desired, which made the raised rivet and panel line detail a bit of a challenge especially when cleaning up the join lines etc. Despite these problems the kit was well regarded and made up in to a good replica.

In 1993 an additional mould tool was made for the kit, and Airfix released an alternative version labelled as a 'B.III Special Dambuster' for the 50th Anniversary of the Dams Raid. Essentially identical to the original B.I/B.III kit, an additional sprue was included with a modified bomb bay and the 'Upkeep' bomb and associated fittings. Unfortunately the 'Upkeep' bomb was moulded with what appeared to represent the wooden shuttering initially applied around the bomb but in the event not used on the operation.

Left and above: Another Airfix second moulding kit, this time built as a modified Lancaster B.1, WU-28, (ex-TW648) for the *Aéronavale* in the early 1950s, in the overall dark blue paint scheme with Xtradecal decals for *Flotille* 25.F, circa 1958. The kit has also been fitted with Freightdog resin wheels, and again, the search windows were cut out of the rear fuselage.
Model by Tony O'Toole

Top and above: A Lancaster B.1(FE), again made from the second moulding Airfix kit, finished in the markings of TW900 'EM-F' of No.207 Squadron, based at RAF Stradishall in 1947. A White Ensign set was used to improve the cockpit interior and undercarriage, (see detail photos). The markings were from an Xtradecal set with the 'M' modified to form the right style. Propellers and exhausts were 'borrowed' from a Revell kit. *Model by Paul Hughes*

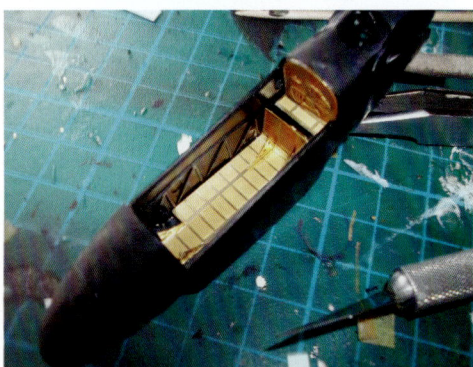

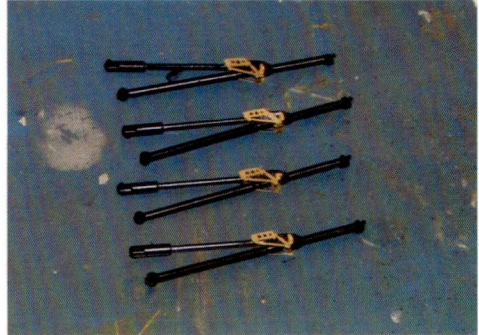

Avro Lancaster B.I/B.II/B.III Special (third moulding)

In 2013, Airfix released a brand new tooling 'multi-variant' kit from which they have produced an initial three boxings. The parts are vastly superior to their predecessors and have finely engraved panel lines, well moulded parts and no half-man/legless machine gunners for the turrets!

Avro Lancaster B.III (Special) The Dambusters

The first variant to be released was the B.III (Special) 'Dambuster', quite obviously linked to the 70th anniversary of the Dams Raid, and quite rightly calculated as an acceptable sales pitch, as any modeller impatient for a standard bomber would probably buy the Dambuster version anyway.

The six sprues, comprising well over a hundred parts, are moulded in grey plastic with one clear sprue, and decals for two machines. One of the sprues is dedicated to this Dambuster version, (which is replaced by a more conventional payload in the B.I/B.III boxing – see later). There is plenty of interior detail, some of which will hardly be seen, especially in the mainwheel wells. Some parts are very finely moulded and the plastic Airfix uses is easy to work with. The transparencies are very well done with delicate framing and good clarity.

The wing panel lines are based on the panel skins and not rivet lines, so the wings do not have the patchwork quilt appearance of the Hasegawa and new Revell kits. The kit is engineered in a slightly different way to most other manufacturers' Lancaster kits because the wings must be able to accept both the B.I/B.III's in-line Merlin and the B.II's Hercules radial engine options. These differ in the way the nacelles blend in with the wing and there are cut-outs in the wing leading edges to allow for this. The other unusual aspect is the way the two wing spars, having been inserted through the fuselage, then have the mainwheel well interiors attached directly to them. This requires that the wings be assembled around the mainwheel wells and onto the spars.

A reasonably detailed and realistic interior is provided. There is a cockpit floor formed by the forward section of the bomb-bay, with the raised pilot's area as a separate part, and a nicely reproduced pilot's seat. There is a navigator's table, for which there is a decal map, partial bulkhead and chair, a flight-engineer's folding seat and instrument panel, again represented in decal form, as is the pilot's instrument panel, which features a throttle box and rudder pedals. The cockpit side walls, including those in the bomb-aimer's area, have integrally moulded detail. There is also a conventional bombsight that needs to be left out, as it was not used in the Dams Raid. Overall, the cockpit as supplied will look quite adequate, especially if seat straps are added.

There are a number of small fuselage windows to add, which some later Lancasters had blanked off. The bomb-bay interior is also nicely detailed, and includes the sidewalls in this area as well as finely moulded bomb crutches (not used with the Dambuster version). There is also the choice of a blanking plate or a Vickers 'K' gun to fit where the FN.64 ventral turret would otherwise be mounted (which is supplied with the B.II's sprues). Airfix supply a clear transparency to insert in the rear fuselage underside to represent the three signalling lights.

The wings include well detailed flaps that can be modelled 'raised' or 'lowered', with two different inboard nacelle end-caps to suit 'raised' or 'lowered' flaps. The wingtip navigation lights are moulded solid, and ideally need cutting out and replacing with clear sprue sanded to shape, but fine aileron actuators are provided and there are cable-cutters for the leading edges.

The engine nacelles are well done, and capture the subtle lines of the real ones. The propeller blades for the Dambuster option

The three current boxings of the latest (third moulding) Airfix Lancaster kit, initially released in 2013.

are the 'narrow' type, but both these and the 'broader paddle blade' type are supplied. The mainwheel wells are very nicely detailed. The open wing rib truss arrangement either side of the wells will benefit from being blanked off with plastic card and painted reddish-brown to represent the internal wing fuel tanks. The undercarriage is also captured well and the mainwheels have the smooth-tread tyres widely used during the Second World War. The tailwheel is a two-piece hub and tyre to enable the anti-shimmy dual ridge to be represented, and fits in to a separate tailwheel leg and fork.

All the glazing is nice and clear with fine framing. The main canopy looks good and the gun turrets are well done with sufficient internal structure represented to satisfy most modellers. The machine gun barrels are also pretty good given the moulding limitations.

A Lancaster B.I, made from the recent third moulding Airfix kit parts, and finished as W4964 'WS-J' and named 'Johnny Walker – Still Going Strong' (based on a period advert for Johnny Walker malt whisky), from No.9 Squadron as it looked prior to Operation *Paravane*, the *Tirpitz* raid of 15 September 1944, carrying a 12,000 lb 'Tallboy' bomb. The new Airfix Lancaster has quite rightly received rave reviews, and now Freightdog have allowed another important variant to be built from these moulds by releasing a resin conversion for a 'Tallboy' bomber. The conversion comes complete with bulged bomb doors, a 12,000 lb Tallboy bomb and the SABS bomb sight.
Model by Tony O'Toole

As Dambuster Lancasters flew without a mid-upper turret a blanking plate is included to fit in its place. Also supplied are the bomb bay's fore and aft fairings and the v-shaped arms that held the Upkeep mine. Airfix also supply the hydraulic motor and chain drive used to spin the bomb, and of course the weapon itself, which, unlike the earlier Airfix Dambuster kit, has a smooth casing without the wooden shuttering. The spotlights used to fix altitude when flying low over water are also included. A nice addition with this kit is a bomb trailer for the Upkeep mine, which appears to be well executed, with weighted tyres befitting its load.

The decals include markings for two aircraft – Joe McCarthy's ED825 'AJ-T' and Robert Barlow's ED927 'AJ-E', which was unfortunately lost during the raid.

Avro Lancaster B.II
Modellers have had to wait over half a century for an injection-moulded kit of the Hercules-engined B.II. Granted, resin conversion sets for the B.II by Paragon and CMR have been available, but this is the first mainstream 1/72 scale kit. I think it is safe to say that the wait has been worth it as Airfix have produced a very good kit.

Based upon the company's basic 'multi-variant' mouldings that are currently considered to be the best overall Lancaster kit in 1/72 scale, Airfix have done a good job of the B.II's Hercules radial engines and their associated features. One large sprue caters for the Bristol Hercules radials, their propellers, cowlings and nacelles. Also included are the bulged bomb-bay doors often associated with the B.II and the FN.64 ventral turret initially carried by some aircraft. These parts are produced to the same high standards as the rest of the kit.

Overall, the cowlings are well done, and the carburettor and oil-cooler air intakes look fine. The hedgehog exhausts are cleverly moulded in three parts and look most convincing, and better than previous injection-moulded examples. The air intakes and exhausts are the standard later pattern, rather than the style fitted to early B.IIs. The propellers and spinners also look good.

The kit's optional bulged bomb doors, which were fitted to most B.IIs, can be displayed open if wished. The FN.64 ventral turret is well done, although many B.IIs weren't fitted with it as it was essentially useless and merely considered to be excess weight. A blanking cover provides for those options that do not require it. Airfix provide the early style shallow bomb-aimer's transparency suited to this version and identify this accordingly in the instructions.

Decal choices feature 'JI-F' 'Fanny Ferkin II' of No.514 Squadron, No.3 Group, RAF Waterbeach Cambridgeshire, November 1944 (with bulged bomb-doors and FN.64 turret), and 'EQ-Z' of No.408 (Goose) Squadron, No.6 Group RCAF, RAF Linton-on-Ouse, Yorkshire, July 1944 (with regular bomb-doors and no FN.64 turret).

Avro Lancaster B.I(FE)/B.III
The B.III is the latest release of Airfix's 'multi-variant' new tooling, following on from the B.I/B.III Special 'Dambuster' and the B.II with Hercules engines. The kit contains six sprues in pale grey and one sprue of clear injection-moulded plastic, and features parts for the later FN.82 rear turret with twin .5in machine guns, (but no Rose-Rice turret option), 'narrow' and 'paddle-bladed' propellers, shrouded and exposed exhausts, early and later style bomb-aimer's blister, optional observation blisters for the sides of the main canopy, and an optional clear-plastic H2S radome.

The instructions are well printed with, not surprisingly, reference to Humbrol paints throughout the build, and the full-colour markings instruction sheet gives the paint colours and decal placement for the two versions – a B.I(F.E) of No.35 Squadron, RAF Graveley, Cambridgeshire, 1945, in the Tiger Force white upper and black under surface scheme, and a B.III 'Frederick II' flown by the Commanding Officer of No.57 Squadron, RAF Scampton, August 1943 in the wartime Dark Earth/Dark Green/Night camouflage scheme.

The build begins by spraying all the interior parts with either Interior Green or matt black depending on the part, followed by fitting the two bomb bay bulkheads to the cabin floor and then the two wing spars to this. The pilot's seat is next which was painted and then assembled and fitted to the cockpit floor, (there is a pilot figure supplied for those who like to crew their models). The cockpit floor was then glued to the cabin floor. The main instrument panel is plain plastic with decals for the instruments. Eduard have released a complete upgrade for this kit's interior should any modeller wish to purchase one, but this kit was built straight from the box.

The cabin floor assembly then slots in to place in the port fuselage half. There are clear parts for the fuselage windows but I did not use these preferring to represent them with Micro Kristal Klear later on – which saves masking them all out. The bombsight was fitted in to the nose and then the navigator's desk, which is just visible through the cockpit glazing when the model is finished, was assembled complete with maps, again supplied as a decal. The radio is catered for by another decal, and yet another decal is supplied for the flight engineer's panel on the starboard side, and then the two fuselage halves were joined together.

Throughout the early stages of the build there are two options to choose from. Option A is for the B.I and Option B is for the B.III. I chose Option B.

The main landing gear ribs were now glued to the main spars and look very impressive when assembled. The aft gear bay bulkheads then fit to the spars and the upper wing surfaces are glued to the spars. The fit is excellent with no filler required. The lower wing surfaces were then glued to the upper wing halves, which are another good fit. The tailplanes were assembled and fitted to the fuselage. These slot and lock together inside the fuselage and provide a really positive fit.

The engines are next. The radiator rear parts were all assembled, painted and then fitted to one half of each engine cowling and then the other halves were joined to them. There is a choice of exhausts, the unshrouded exposed manifold type and the flame-damping shrouded type. These were fitted along with the intakes on all four engines and then the engine assemblies were fitted to the wings. When fitted there is a slight gap between the wing and engine at the top, which may be a legacy from the wing moulds being utilised for the B.II.

The flaps were assembled and put to one side. They can either be fitted lowered or raised. The fins were assembled and fitted to the tailplanes at this stage. Bomb brackets were fitted in to the bomb bay and then the machine gun turrets were painted and assembled.

A lengthy period of masking now began, starting with the cockpit canopy and then the gun turret framework. The canopy was glued in to position and the gun turrets placed in to their respective locations, following which the model was sprayed with primer. Any gaps were dealt with, (there were very few), and then the model was ready for painting.

After painting, (in this instance I chose the wartime Bomber Command Scheme Dark Earth/Dark Green/Night option), all the masking was removed and final assembly began by fitting the undercarriage, which is well modelled and looks like the real thing. The propellers were painted, assembled and fitted to the engines and then finally the aerials were fitted and the model was complete.

Having built several of Airfix's 'second moulding' Lancasters a few years ago, the arrival of this new tooling is most welcome and long overdue, especially the radial-engined B.II released last year.

This range of Airfix Lancaster kits, which are generally considered to be the best 1/72 scale Lancaster kits so far released, are well moulded, the slightly exaggerated fabric effect on ailerons notwithstanding, with engraved panel lines (and not a rivet in sight!), accurately shaped, have good interior detail, and provide the basis for several Marks, including post-war variants.

The parts assemble well, the instructions are good and the decals likewise. There are no bombs supplied in the B.I/B.III kit, (although bombs are available in a separate kit, the 'WW2 Bomber Re-supply' set which is well worth purchasing). Airfix seem to have captured the key shapes better than their competitors and have the best lines and equal best internal detail. Surface detail is a little bit heavier than some of its rivals, but this is balanced by there being less of it. A thoroughly enjoyable build, and for overall value, a clear winner.

■ Assessment by Andy McCabe

FROG

1/72 scale

Avro Lancaster B.I

Released in 1976, a couple of years before the second Airfix moulding, and just a few months before the company sadly folded, it was product of its time, and featured raised panel lines and very basic interior detail, but thankfully no excessive rivet detail.

The main 'plus' with this kit was that it featured bulged bomb bay doors and a 12,000 lb 'Tallboy' – the first time either had been kitted in any scale let alone 1/72. This Lancaster was generally fairly accurate, but had an inaccurately-shaped mid-upper turret and the wrong style of tailwheel. It failed to capture the engine cowling outlines correctly nor the way they blended into the leading edges, giving the appearance of being slung far too low.

It did feature separate elevators, but no other separate control surfaces, and had perhaps the most restrained surface detail of any of the 20th Century Lancaster kits, all of which is relatively academic as the kit is no longer available and in any case it has been superseded by better kits.

Decals provided markings for NG494 'KC-B' of No.617 Squadron, May 1945 and LM220 'WS-Y' of No.9 Squadron, April 1945.

The 1976 Frog 1/72 scale kit box top and decal sheet. The main 'plus' with this kit was that it featured bulged bomb bay doors and a 12,000 lb 'Tallboy'. The decals provided markings for NG494 'KC-B' of No.617 Sqn, May 1945 and LM220 'WS-Y' of No.9 Squadron, April 1945.

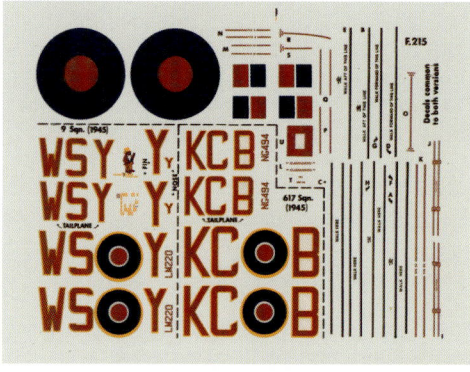

MATCHBOX
1/72 scale

Avro Lancaster B.I/III
The Matchbox Lancaster was released around late 1979/early 1980 – presumably with the younger modeller in mind. The four sprues were presented in Matchbox's distinctive different coloured plastic – in this instance one each in rather appropriate brown and green and two in black.

Panel line detail was engraved on the wings and tail, but raised on the fuselage and engine nacelles, and the fuselage nose section was separate, which made for an awkward join to sand smooth. Interior detail was very basic, but the tailwheel was the correct Marstrand anti-shimmy style.

Various options were included in the kit, such as bulged or standard bomb bay doors, (but no Tallboy, or other types of bomb); a mid-upper turret with a separate surround or dorsal blanking plate; a solid plastic H2S radome or smooth under fuselage section.

Decal options were for R5677 'ZN-A' of No.106 Squadron, September 1942, with strangely proportioned fuselage roundels; NG206 'WS-J' of No.9 Squadron, December1944; and an RAAF example, A66-1, based at Tocumwal, NSW in 1945.

Like the Frog offering, this was a typical late 1970s/early 1980s product with the usual characteristics of the brand. It was

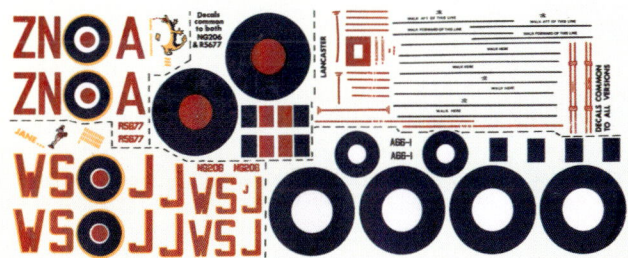

generally well received when first released and was regarded as being more accurate than the Frog kit, but again this is academic as it is no longer available nor indeed in production.

This kit was released around late 1979/early 1980, moulded in the company's distinctive different-coloured plastic. Various options were included in the kit such as bulged or standard bomb bay doors, mid-upper turret or blanking plate and a H2S radome or smooth under fuselage section. Decals were for R5677 'ZN-A', No.106 Sqn, September 1942; NG206 'WS-J' of No.9 Sqn, December 1944; and A66-1, RAAF based at Tocumwal, NSW in 1945.

HASEGAWA
1/72 scale

Hasegawa released the first modern, engraved detail, kit of the Lancaster in 2005. This is a good kit, typical of the brand, with clean crisp mouldings and fine surface detail. However, it suffers from indifferent engine radiators, incorrectly placed cockpit canopy escape hatch and an oversized tailwheel. The fit of the engine nacelles is not as good as it could be and the interior detail is rather basic. The number of panel lines is excessive – every rivet line being represented by an engraved line, as opposed to a sheet of metal forming a panel. This makes the wing surfaces look a bit like a patchwork quilt. True to form, Hasegawa have subsequently released numerous boxings, offering 'Tallboy', 'Grand Slam', Dambuster and lifeboat carrying variants.

Avro Lancaster B.I/B.III
Hasegawa's Lancaster was released just before the 'new' Revell kit (see later) and includes a 'post-war Lancaster' boxing which offers additional parts not found in the standard kit, such as Lincoln-style fins, a twin .5in mg FN.82 rear turret, bulged bomb bay doors and a flare and strike camera mounting as was fitted to most (but not all) maritime reconnaissance Lancasters below the rear turret.

It also includes decals for three aircraft including a Tiger Force scheme bomber and two Coastal Command Lancasters, one of which is overall Dark Sea Grey and the other in the earlier Medium Sea Grey and White scheme, making this kit more relevant to the post-war focus of this book. Unfortunately, this kit is over twice the price of the Airfix or Revell kits in the UK!

Upon initial inspection the parts look crisp and the mainwheels and tailwheel are the correct styles. The detail is relatively simple, and the roof of the bomb bay, which also serves as the cockpit floor, is covered with large holes which need to be filled in – not really acceptable in a kit costing so much money!

Positions are provided for the navigator and wireless operator but they are composed of an over-simplified one-piece part. A separate GEE box is provided and decals are provided to give some representation of the radio and navigational equipment. Decals are provided for the engineer's panel on the fuselage wall and also the pilot's instrument panel, (which actually represents the instrument panel of a twin-engined Manchester and not a four-engined Lancaster!) – the engineer's panel looks wrong too.

However, neither is obvious and shouldn't be too noticeable through the glazing of a finished model. The cockpit interior isn't as good as that of the Revell kit, although hardly any of the detail can be seen through the canopy anyway – and the Hasegawa interior looks fine once the canopy is in place, which is what really matters.

As with the Revell kit, there is no bombsight provided, nor the distinctive projections that appear in the nose bubble on either side of it, but Hasegawa do provide the option of cutting out and fitting the ASR/GR/MR windows at the rear of the fuselage.

This model also includes spars for the main and tailplanes that fit through the fuselage which makes attaching the flying surfaces so much easier. Once the wing spar has been attached to the roof of the bomb bay/cockpit floor and the tail spar is also in position, the fuselage halves can be joined. The fit is superb. The kit also includes the row of windows in the rear fuselage, but this time they are fitted individually from outside which can aid the painting process if a version using these windows is chosen.

The wings also come as a pair of upper and lower halves for each side, without any separate control surfaces, but with superb panel line detail. The mainwheel well interiors are not as detailed as some other 1/72 scale kits but are moulded into the lower wing halves easing construction.

The engine nacelles are easy to construct and are crisply detailed. They include radiator grilles that can be added from the front after the model has been painted, making painting so much easier as the radiators do not need to be masked off.

The nacelles were easy to attach and fit well. When the wings are attached to the fuselage the model feels much more rigid. The tailplanes can be attached at this stage with a choice of the rounded Lancaster-style, or, Lincoln-style 'endplate' rudders. The kit provides decals for TW657 'TL-C', which is the same 'Tiger Force' scheme Lancaster that appears on Xtradecal sheet X72061.

Also included is an air intake above the starboard wing root which was part of the (FE) conversion and is vital for reproducing most post-war Lancasters. The front turret doesn't have a rear section to it, but once it is in place on the model this is not apparent and it has a more convincing shape than Revell's offering. As mentioned, the Hasegawa kit also includes a twin .5in mg FN.82 tail turret.

The main cockpit canopy shape is good and includes optional astrodomes and side windows, with or without blisters, although the pilot's escape hatch in the roof is in the wrong place. This may not seem like such a big problem but it does make the framework look very odd.

The bomb load consists of eighteen very good 500 lb bombs (wrongly described in the instructions as 200 lb bombs – albeit no such type existed in British service – so presumably a typo), and a 4,000 lb 'Cookie', with individual bomb racks to mount them on. The tailwheel is the correct anti-shimmy type with a groove down the middle, and a clear H2S radome is included. Both air intake options, the short style and the long tropicalized style, on either side of each engine nacelle, are also included.

The main undercarriage detail is good and the mainwheels themselves have the correct hub detail and even have 'weighted' tyres of the treadless wartime type. The kit includes 'Rebecca' radar receivers on either side of the forward fuselage and parts are provided for the windscreen anti-icing nozzles, located in front of the windscreen. To fit the wingtip navigation lights, the appropriate sections have to be cut away using a razor saw but the clear parts are too big and need to be cut down using a craft knife before being attached.

■ Assessment and model by Tony O'Toole

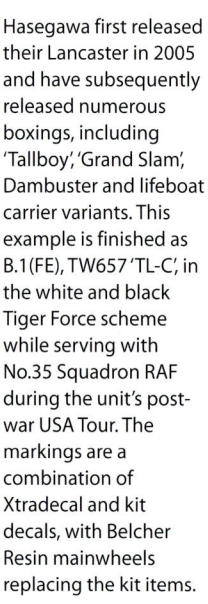

Hasegawa first released their Lancaster in 2005 and have subsequently released numerous boxings, including 'Tallboy', 'Grand Slam', Dambuster and lifeboat carrier variants. This example is finished as B.1(FE), TW657 'TL-C', in the white and black Tiger Force scheme while serving with No.35 Squadron RAF during the unit's post-war USA Tour. The markings are a combination of Xtradecal and kit decals, with Belcher Resin mainwheels replacing the kit items.

REVELL 1/72 scale

The 'old' Revell Lancaster kit sprues which were released with the company's Spitfire and Hurricane in a Battle of Britain presentation box. The kit featured Clydeside-style metal plating surface detail and had accuracy and outline issues, (both the standard bomber and Dambuster were based upon the same common tooling).

The painting instructions for the original tooling Revell Lancaster.

Avro Lancaster B.I and 'Dambuster' Special (original tooling)

In 1964 Revell released two versions of the Lancaster in 1/72 scale – a standard B.I bomber and a Dambuster. Originally moulded in black shiny plastic, both kits featured Clydeside-style metal plating surface detail, but were sturdy models when completed and easy to build. Unfortunately, both had serious accuracy and outline issues, (they were based upon the same common tooling), and are now best considered, like the original Airfix kit, as collectable items.

AVRO LANCASTER MK-I

H207 AVRO LANCASTER MK I

The exploits of the English Lancaster bombers during World War II have become legendary. Lancasters were employed in the dramatic "Dam Buster" raids and destroyed the dreaded German battleship Tirpitz.

A MONUMENT TO GREATNESS

One of the most famous Lancasters of the Royal Air Force now forms part of a permanent monument in the Royal Air Force Museum at Hendon, England, known by her code name 'S' for "Sugar". This Lancaster stands as a tribute to these great airplanes and those who flew them.

THE QUEEN GOES TO WAR

Sugar entered service as "Q" for "Queenie", and she joined No. 83 Squadron at Scampton in the summer of 1942. Her first operational flight was made in a bombing attack on Wilhemshaven on July 8, 1942. On July 11, she participated in the famous dusk raid on Danzig, the most distant target yet to be hit by the Bomber Command. Her first trip to Italy came on November 6, and the first of many night visits to Berlin was made on January 16, 1943.
Late in 1943, after being thoroughly overhauled, Queenie joined No. 467 Squadron, Royal Australian Air Force in Waddington, Lincolnshire, and became "S" for "Sugar".

FOOTBALL IN THE SKY

On November 26, 1943, Sugar was returning from a raid on Berlin when she met with near disaster, but the ruggedness typical of the Lancaster brought her safely home. The operations diary records the incident:
"Flying Officer J. A. Colpus tried Aussie-rules-football with another Lancaster and tried to bump it out of the sky. The aircraft went into a severe dive to port, but by applying full rudder and aileron trim, the aircraft straightened, but it still needed a lot of pressure on both rudder pedals and the control column to maintain height. The aircraft was our old reliable "S" for "Sugar", and it had completed 80 trips. In this kite the pilot and navigator go to sleep coming home, for it knows its own way back from almost any any target".

100 FOR SUGAR

Sugar completed her "century" flight on a raid on Bourg Leopold in Belgium, but she was forced to fight for her life to gain the honor. In the intense activity over the target, two JU 88 night fighters made ten determined assaults on the venerable Sugar. Her crew, through teamwork and close coordination, successfully evaded the attackers and Sugar escaped untouched. As she rolled to a stop at her home base she was met by the rousing cheers of many officers and airmen who waited up to greet the veteran.
Sugar's last operational sortie was made on April 23, 1945 in a flight to Flensburg, but the target was obscured by heavy cloud and the plane was unable to unload her bombs.
Today she proudly displays her 137 miniature bombs and D.F.C. and D.S.O. emblems in defiance of the slogan emblazoned on her nose. " 'No enemy plane will fly over the Reich Territory' – Herman Goering".
The Avro Lancaster MK I was powered by four 1,640 hp Rolls Royce Merlin engines. The wingspan and length were 102 feet and 69'6", respectively. Overall height was 20 feet. Maximum speed of the Lancaster was 287 mph, service ceiling was 24,500 feet. Two 0.303 machine guns each were located in nose, dorsal and tail turrets.

BUILD QUEENIE OR SUGAR

Today Sugar rests in the Royal Air Force Museum at Hendon after standing for a number of years at the entrance to R.A.F. Scampton in Lincolnshire. The Revell model of this historic aircraft is provided with two sets of code letters to enable you to build her as either Queenie or Sugar.

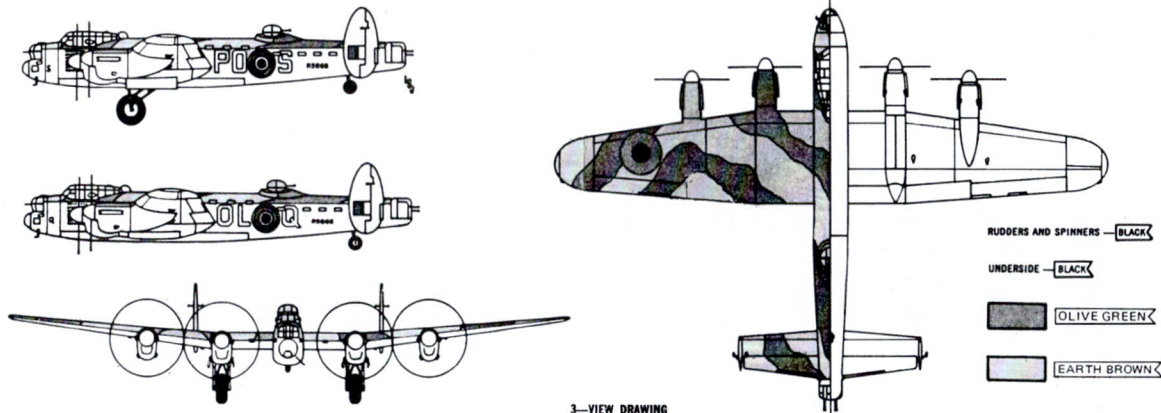

3—VIEW DRAWING

A period advert for the Revell Lancaster kits when they were first released in 1964.

The 'old' Revell Dambuster Lancaster box top and decal sheet with markings for Wg Cdr Guy Gibson's aircraft.

The new Revell Lancaster B.I/B.III box top illustration, featuring B.III, LM739 'HW-Z squared' of No.100 Squadron RAF. The other decal option is B.I, W4964 'WS-J' of No.9 Squadron RAF.

Avro Lancaster B.I/B.III (new tooling)

Revell released their new tool 1/72 scale Lancaster B.I/B.III kit in early 2008. The panel lines are nicely engraved, the cockpit canopy is the correct shape and the engines can be built with the exhaust manifolds exposed. The interior of the cockpit is excellent with the provision of optional padded or unpadded seats for the pilot. The inclusion of crew positions for the navigator and wireless operator is also a nice touch whilst the rib detail on the inside walls of the fuselage looks really effective. Behind the wireless operator's position there is a large box section which represents the famous Lancaster 'wing spar', over which a generation of British and Commonwealth bomber crews have had to scramble, and this sticks out through openings in the fuselage sides to provide a very clever way of maintaining the dihedral of the wings. The bomb bay is also nicely done with separate bomb racks and the detail inside the mainwheel wells is superb with separate wing spars and webs to fit.

Decals are provided for the instrument panel, Flight Engineer's panel, wireless sets, Gee box and even some seat straps, which look quite effective when viewed through the cockpit glazing. The detail extends in to the bomb aimer's position in the nose, which makes it a 'busy looking' area, but for some reason the bomb sight has been omitted, along with the prominent projections found on either side which can be seen from the exterior through the nose 'bubble'. The yellow painted hand rail leading from the cockpit is also missing. Even with these omissions this area is still much better than many other Lancaster kits.

The front, mid-upper and rear turret guns are a little bit basic and disappointing. Revell have included an alternative blanking plate for the mid-upper turret station and one for the little-used ventral turret which fit well and can be slotted into place without the need for filler. Revell have also included the air intake that provided air conditioning to either the crew or the avionics when operating in tropical conditions, located on the fuselage over the starboard wing root and fitted to practically every post-war Lancaster.

The wings are simplicity themselves. The main parts consist of an upper and lower section for each wing, with no gimmicks such as moving control surfaces included. This also applies to the horizontal tailplane parts too. Inside each wing there is a detailed mainwheel well bay that can be built up, with a prominent looking set of wing ribs on either side, and a pair of authentic-looking webs join others already moulded into the roof. Fully detailed bulkheads are included too which slot into place followed by the large castings that will hold the undercarriage legs.

The inner engine nacelles also contain a very good representation of a Merlin engine. All the engine nacelles are relatively easy to assemble and the exhausts can be built up with or without the anti-glare shrouds, mak-

The new Revell Lancaster B.I/B.III kit finished in the later scheme carried by the few remaining Lancaster MR.3s operated by the School of Maritime Reconnaissance at St Mawgan in the mid-1950s, namely overall Dark Sea Grey. The markings are from an Xtradecal sheet, with new antennae under nose and camera/flare housing from a Hasegawa kit attached to tail. Once again, search windows were cut out of the rear fuselage.
Model by Tony O'Toole

Left: The new Revell Dambuster box top illustration, again featuring the markings for Wg Cdr Guy Gibson's aircraft.

Below: Another Revell Lancaster B.I/B.III tooling, this time finished as TX269 'RL-N' of No.37 Squadron, based at Ein Shimar, Palestine in 1946/47. The model is painted in the Extra Dark Sea Grey/Dark Slate Grey/Sky scheme with markings coming from a combination of RAFDec decals and the spares box. New antennae was fitted under the nose, with home-made resin mainwheels, and again, search windows were cut out of rear fuselage.

ing it possible to model an aircraft which had the exhaust manifolds exposed, although they seem to stick out further than those on other manufacturers' kits. One part of the nacelles which are a little sub-standard are the carburettor air intakes on either side which are quite 'box like'.

The kit's decal options are for Lancasters fitted with the 'narrow' type of propeller blades, but Revell have also provided optional 'paddle-bladed' propellers. The completed engine nacelles are designed to be attached to the completed wings and the fit is mostly good although the left inner nacelle needed to be held in place using clamps until the glue had fully dried.

The main undercarriage legs and their supporting struts have plenty of detail, but the smooth, untreaded tyre, mainwheels are far too big and slab-sided, whilst the hub sections bear no actual resemblance to any Lancaster mainwheel! These can be replaced by the excellent resin treaded wheels from Belcher Bits, a Canadian resin accessory company.

Once the fuselage and wings are complete, the wings simply slot onto the centre section box. At the time of the kit's release there was much talk amongst modellers about the dihedral of the wings, especially the outboard sections. At first glance, the dihedral does look a little 'flat', but it is in fact correct. When the aircraft is on the ground the dihedral is not as marked as when it is in the air – which is possibly where the confusion arose.

The turrets can be assembled and fitted after the fuselage halves are joined, but the shape of the nose turret is a bit suspect and is missing its vertical frame line! However the rear turret and canopy are more than adequate, the canopy provided the option of the tall or short astrodomes and flat or bulged side windows.

Two types of bomb aimer's blister are included – the shorter early style and the later, more bulged style first used by the B.III, with the two Z-Type IFF Transponders which were used to identify the aircraft as friendly to the rear-warning radars on other aircraft. Further options include the oval or rectangular downwards viewing windows behind the nose blister.

Decal options included B.I, W4964 'WS-J', of No.9 Squadron RAF and B.III, LM739 'HW-Z squared', of No.100 Squadron RAF. The Revell kit featured here is finished as Lancaster ASR.3, TX269 'RL-N', of No 38 Squadron, Palestine, 1946, in the Temperate Sea Scheme.

■ Assessment and models by Tony O'Toole

TAMIYA

1/48 scale

Avro Lancaster B.I/B.III

Originally released in 1976, this kit was subsequently re-released, initially as a 'limited edition' version, in 2009. It included electric motors (to spin the propellers) and pre-painted canopy and turret frames: this kit was released again in early 2012 with some additional parts.

For the current (2012) release, which is still available at the time of writing, the electric motors have been deleted as have the pre-painted versions of the clear parts. The 2009 'limited edition' release also provided some additional parts including optional engine cowlings with recessed panel detail and the option of unshrouded exhausts, 'narrow' propeller blades (as well as the original 'paddle' bladed propellers) and bulged and 'weighted' mainwheels treadless tyres, all of which are included in the current (2012) release. This release also features two new hemispherical bomb aimer's domes, both larger than the early production type included in the original kit boxing – one is plain, while the other has the two infra-red light transmitters for the 'Z' equipment and parts for the later FN.82 rear turret armed with two .5in machine guns.

All the solid parts are now moulded in pale grey plastic instead of the original's black, but the surface detailing is still raised, like the original, although it is quite fine and not overdone. The model retains its fairly basic interior, and the pilot's area which is visible through the canopy glazing could really do with a bit of attention. The bomb aimer's station is also a bit sparse as are the gun turret interiors. There is also an error in the kit's positioning of the navigator's and radio operator's table and associated seats, which are positioned too far forward, that has a 'knock-on' effect on the two square windows on either side of the forward fuselage. Ideally, these square windows need moving back to a position under the small astrodome at the rear of the cockpit glazing.

Keeping with the aircraft's interior, the instructions advise you to glue the pilot's parachute in the pilot's seat pan. However, all the crew had places to store their parachutes and the pilot's would never be left in the seat pan. In fact the only one that can be readily seen from the outside of the airframe is on the bulkhead behind the bomb aimer's position.

The forward section of the fuselage interior including the bomb aimer's station and the flight deck was painted matt black, including the pilot's seat, on which the yellow disc on the armour plate headrest should be painted on both sides. The navigator's/radio operator's table appears to have been grey/green with black seats. The rest of the interior from the main wing spar, which is right behind the radio operator's station, to the rear turret is interior grey/green. For modelling purposes, the interior black can be painted a very dark matt grey, allowing the interior to be slightly more visible through the cockpit canopy, and allowing other 'black' items to be painted various tones of semi-gloss and gloss black, to highlight them a little and give a more 'scale' appearance. All seat cushions, back rests and padded armrests were either black or dark glossy green leather. The seat cushions in the kit have deep ribbed gaps which is incorrect as they were smooth leather, so the gaps should be filled.

If assembling the original kit's four gun FN.20 rear turret, it should be noted that the machine guns and main mounting brackets are shown back to front on the instruction sheet with the mounting brackets sloping in the wrong direction. The left one should be on the right and the right-hand one on the left!

A new large decal sheet with four subjects, B.III, ED888 'PM-M', of No.103 Squadron; B.III, ED905 'BQ-F', *Ad Extremum* of No.550 Squadron; B.I, NG347 'QB-P', 'Picadilly Princess' of No.424 Squadron; and B.1(FE), TW880 'TL-F' of No.35 Squadron in the white and black Tiger Force scheme.

Despite being over three decades old, Tamiya have kept the basic moulds in excellent condition and, by enhancing the kit with the new-tool additions, they have ensured that their Lancaster is still a viable modelling proposition in the 21st Century, and it still remains the only injection-moulded Lancaster ever produced to 1/48 scale.

The 2012 release box top, which includes two new hemispherical bomb aimer's domes – one which is plain and the other with the two infra-red light transmitters for the 'Z' equipment – the original 'paddle' blade and new 'needle' blade propellers, the later Fraser Nash FN.82 rear turret, improved engine nacelle tops and 'exposed' exhaust manifolds. A new large decal sheet is also included with four subjects – B.III, ED888 'PM-M' of No.103 Squadron; B.III, ED905 'BQ-F' *Ad Extremum* of No.550 Squadron; B.I, NG347 'QB-P', 'Picadilly Princess' of No.424 Squadron; and B.1(FE), TW880 'TL-F' of No.35 Squadron.

'Dambuster/Grand Slam Bomber' boxing

Tamiya released a Type 464 Provisioning Lancaster B.III, as used in the Dams Raid, at the same time as the standard B.I/B.III in 1976, which was a slightly revised version of the B.I/B.III but with parts for a new 'open' bomb bay and optional parts for the 'Upkeep' mine or a 22,000 lb 'Grand Slam' bomb.

Unfortunately, these Type 464 Provisioning B.III Specials were fitted with the larger, clear, hemispherical bomb aimer's dome, which only became available with the 2009/2012 Tamiya boxings. Most of the Operation *Chastise* aircraft were also fitted with VHF radio sets to enable the aircraft to communicate with each other better, especially at low level, which can be identified by the fitting of a thin blade VHF aerial under the starboard side of the nose forward of the port side camera ports.

No indication of the spotlights that were used to determine the height of the aircraft during the Dambuster bombing runs is included in the kit or in the instructions. In fact, the front spotlight was positioned in the lower port side camera position, to the rear of the bomb aimer's window and the rear light was positioned in the bomb bay, on the angled bulkhead immediately to the rear of the 'Upkeep' bomb. Both spotlights shone their beam through short tubes, angled slightly forward and to starboard, so that they could be seen by the Flight Engineer looking through the starboard cockpit teardrop fairing, forming a figure '8' on the surface of the water when the aircraft was at 60 feet. So, holes need to be drilled in the position for the lower camera port and one in the centre of the angled bulkhead immediately to the rear of the 'Upkeep' bomb, and short lengths of tube glued in to them. A blanking plate for the deleted mid-upper turret is provided.

Only the broad 'paddle blade' propellers are included in the kit, but both the Operation *Chastise* and 'Grand Slam' were fitted with the narrow chord 'needle' propeller blades. The Tamiya kit's engine nacelles are possibly the worst bit of this kit. Not only are they a bit suspect in shape, but they are a rather poor fit to the main nacelle fairings and wing leading edges. The side-mounted carburettor air intakes also look too small. Belcher Bits No.15 'Lancaster Engines' resin replacement engine nacelles, which are beautifully cast in grey resin, are recommended replacements for the kit items, and come with alternative exhaust manifolds and optional flame damping shrouds. Underside coolant and oil radiators air outlet doors are also included, which are often seen 'open' on parked Lancasters.

The Tamiya mainwheel undercarriage legs are good, but this Dambuster/Grand Slam boxing kit only includes late production treaded tyres. So, Belcher Bits to the rescue again, with their Set No.12, 'Lancaster Smooth Tread Bulged Tyres'. It would also appear that the tail wheels on the Type 464 Provisioning Lancasters were 'smooth' without the anti-shimmy groove around the centre of the tyre as moulded by Tamiya.

Decals for five Dambuster aircraft, including Gibson's ED932 'AJ-G', plus two 'YZ'-coded Grand Slam machines are included, but the upperwing red/blue roundels in the kit are too small, (they were a massive 100 inches in diameter), so aftermarket options will need to be sourced for these.

Opposite and above: The standard B.III kit painted up to represent Lancaster ASR.3, RF325 'P9-J', of the Air-Sea Warfare Development Unit circa 1948. Made by the author, the model is finished in a well-worn Extra Dark Sea Grey and Dark Slate Grey upper surfaces with Sky under surfaces scheme. The heavy exhaust stains (indicating a well-used machine), ran 'over' the wing on either side of the inboard engines, but only on the inner side of the outboard engines, due to the dihedral of the outer wing panels. However, exhaust staining from all four engines was apparent on the nacelle sides and under the wings.
Model by Martin Derry

The Tamiya Lancaster Mk III 'Dambuster/Grand Slam Bomber' boxing, finished as ED932 'AJ-G' of No.617 Squadron, May 1943, as flown by Wing Commander Guy Gibson and his crew on the night of 16/17 May 1943. The Tamiya mainwheel undercarriage legs are good, but the original boxing kit only includes late production treaded tyres so, Belcher Bits 'Lancaster Smooth Tread Bulged Tyres', were used. It would appear that most of the actual 'Upkeep' bomb/mines were painted in an oxide red colour, although some may have been painted green, and at least one was recorded as being painted black.
Model by Neil Robinson